Notes on HOMOLOGICAL ALGEBRAS

by

JOSEPH J. ROTMAN
University of Illinois, Urbana

VAN NOSTRAND REINHOLD COMPANY
NEW YORK CINCINNATI TORONTO LONDON MELBOURNE

Van Nostrand Reinhold Company Regional Offices:
Cincinnati, New York, Chicago, Millbrae, Dallas

Van Nostrand Reinhold Company Foreign Offices:
London, Toronto, Melbourne

Manufactured in the United States of America

Published by Van Nostrand Reinhold Company
450 West 33rd Street, New York, N.Y. 10001

Published simultaneously in Canada by
D. Van Nostrand Company (Canada), Ltd.

10 9 8 7 6 5 4 3 2 1

PREFACE

These notes are the transcript of my lectures on Homological Algebra given at Urbana in Spring 1968. I have made no effort to be encyclopaedic, and this is merely a leisurely stroll into the subject. There are at least two advantages to this approach. First, Homological Algebra is not yet universally appealing to beginning graduate students, so there is a need to exhibit connections with loveable mathematics. Second, if $T(t)$ is the number of theorems proved at time t, and if $D(t)$ is the number of definitions given at time t, then it is possible to believe that $\lim_{t \to \infty} T(t)/D(t)$ is strictly positive.

It is assumed the reader has been exposed to enough algebra so that he is comfortable with exact sequences, tensor products over noncommutative rings, and, perhaps, direct and inverse limits.

It is a pleasure to acknowledge my thanks to an enthusiastic class who listened to the lectures and kept me alert. My thanks also to the secretarial staff in Urbana, especially Mrs. Norma Stevenson, for an excellent job of typing.

JJR

Urbana, 1968

TABLE OF CONTENTS

ELEMENTARY DEFINITIONS

Here is a brief account of terms not defined in the text.

If R is a ring (always associative with 1), a **left R-module** is an additive abelian group A upon which R acts nicely, i.e., there is a function $R \times A \to A$ with $(r, a) \mapsto ra$ satisfying:

(i) $r(a + a') = ra + ra'$;

(ii) $(r + r')a = ra + r'a$;

(iii) $(rr')a = r(r'a)$;

(iv) $1a = a$,

where $a, a' \in A$ and $r, r', 1 \in R$.

A **right R-module** is an abelian group upon which R acts just as above except that the third identity is replaced by

(iii)′ $(rr')a = r'(ra)$.

If A is a right R-module, one writes $(r, a) \mapsto ar$; with this notation, (iii)′ becomes

(iii)″ $a(rr') = (ar)r'$.

If A and B are both left R-modules, an **R-homomorphism** (or **R-map**) is a function $f\colon A \to B$ with

$$f(a + a') = fa + fa'$$

and

$$f(ra) = rf(a), \text{ all } a, a' \in A, \quad r \in R.$$

There is a similar definition of R-homomorphism between right R-modules (in fact, each "left" definition has its obvious "right" correspondent).

An R-map f is **monic** (or a **monomorphism**) if f is one-to-one; f is **epic** (or an **epimorphism**) if f is onto; f is an **isomorphism** if it is monic and epic.

If A is a left R-module, a **submodule** is a subgroup A' of A with $rA' \subset A'$, all $r \in R$. The **quotient module** A/A' is the quotient group made into an R-module by

$$r(a + A') = ra + A' .$$

If $f\colon A \to B$ is an R-map, then **ker** $f = \{a \in A\colon fa = 0\}$ is a submodule of A and **im** $f = \{b \in B\colon b = fa \text{ for some } a \in A\}$ is a submodule of B. The **Noether isomorphisms** are:

I $A/\ker f \cong \operatorname{im} f$;

II $(A_1 + A_2)/A_2 \cong A_1/(A_1 \cap A_2)$, where $A_1, A_2 \subset A$;

III $A/C \cong (A/A')/(C/A')$, where $A' \subset C \subset A$.

If $f\colon A \to B$, then **coker** $f = B/\operatorname{im} f$.

A sequence of R-maps

$$\cdots \longrightarrow A_{n+1} \xrightarrow{f_{n+1}} A_n \xrightarrow{f_n} A_{n-1} \longrightarrow \cdots$$

is **exact** if $\operatorname{im} f_{n+1} = \ker f_n$ for all n.

A diagram of R-maps

$$\begin{array}{ccc} A & \xrightarrow{\beta} & B \\ {\scriptstyle \alpha'}\downarrow & & \downarrow{\scriptstyle \alpha} \\ C & \xrightarrow{\beta'} & D \end{array}$$

commutes if $\beta'\alpha' = \alpha\beta$. A larger diagram commutes if, given any two vertices, A, B, any two paths (= composites) from A to B are the same map.

If A is a right R-module, B a left R-module, then $A \otimes_R B$ is an abelian group with generators all $a \otimes b$, $a \in A$, $b \in B$, and relations:

$$(a + a') \otimes b = a \otimes b + a' \otimes b \ ;$$

$$a \otimes (b + b') = a \otimes b + a \otimes b' \ ;$$

$$ar \otimes b = a \otimes rb \ .$$

If R is commutative, then $A \otimes_R B$ is an R-module if we set $r(a \otimes b) = ar \otimes b = a \otimes rb$. (If R is not commutative, this definition does not satisfy the associative law.)

The important point about tensor products is that they convert R-bilinear maps to R-linear ones, i.e., given the diagram

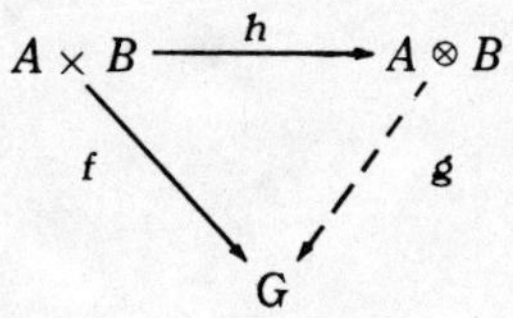

where $h(a, b) = a \otimes b$ and f is R-bilinear, there is a unique map g making the diagram commute. Tensor product is associative (when defined). Also, if $f\colon A \to B$ and $f'\colon A' \to B'$, there is a unique map $A \otimes A' \to B \otimes B'$, denoted $f \otimes f'$, that takes $a \otimes a'$ into $fa \otimes f'a'$. The reader is referred to Bourbaki, *Algèbre Multilinéare,* for details.

HOMOLOGICAL ALGEBRA

§1 *Introduction*

The notion of homology (from the Greek: *homologos*– agreeing, corresponding) arose at the end of the 19th Century. In modern language, one wishes to count, for any integer $n \geq 0$, the number of "independent" n-dimensional holes in a topological space. In the 1940's, it was observed that there are strong analogies between the algebra arising from homology and certain phenomena of classical algebra. For example, the study of extensions of groups leads to formulas quite similar to those arising in topology. This is one aspect of homological algebra: the study of algebraic objects, e.g., groups, rings, by "homological methods." A second aspect of the subject is Categorical Algebra, which we shall not attempt to define (but which is intimately connected with homology also). After this vague preamble, we introduce a language and some examples.

First, a set-theoretical remark. We distinguish between *sets* and *classes;* sets are "small" classes, i.e., a set is a class that has a cardinal number. Thus, the class of all groups is not a set; the class of all countable groups is not a set; the class of isomorphism classes of countable groups is a set (it has cardinal $2^{\aleph_0}$).

DEFINITION. A **category** $\mathcal{C}$ consists of three data subject to certain axioms: (1) a class of **objects** (denoted obj $\mathcal{C}$);

(2) for each ordered pair of objects A, B, a *set* of **morphisms** (denoted $\mathrm{Hom}_{\mathcal{C}}(A, B)$, or $\mathrm{Hom}(A, B)$);

(3) a **law of composition** of morphisms (more on this in a moment).

REMARKS. (i) Hom(A, B) is only required to be a set; in particular, it is possible that Hom$(A, B) = \phi$ for some $A, B \in$ obj $\mathcal{C}$.

(ii) Our motivation is that Hom(A, B) resemble functions from A to B; indeed, one writes $f: A \to B$ for $f \in$ Hom(A, B). However, this is only motivation, and the elements of Hom(A, B) need not be functions (if, for no other reason, that A, B need not be sets!)

We can now say more about the law of composition. Given $A, B, C \in$ obj $\mathcal{C}$, we wish to "compose" morphisms

$$A \xrightarrow{f} B \xrightarrow{g} C$$

to obtain

$$A \xrightarrow{gf} C,$$

i.e., a law of composition is a function

$$\text{Hom}(A, B) \times \text{Hom}(B, C) \to \text{Hom}(A, C)$$

which we write

$$(f, g) \mapsto gf .$$

Here are the axioms:

1. Composition is associative, i.e., in the situation

$$A \xrightarrow{f} B \xrightarrow{g} C \xrightarrow{h} D ,$$

we have $h(gf) = (hg)f$.

2. There are **identity morphisms** i.e., for each object A, there is a morphism $1_A \in$ Hom(A, A) (so this Hom is nonempty!) satisfying

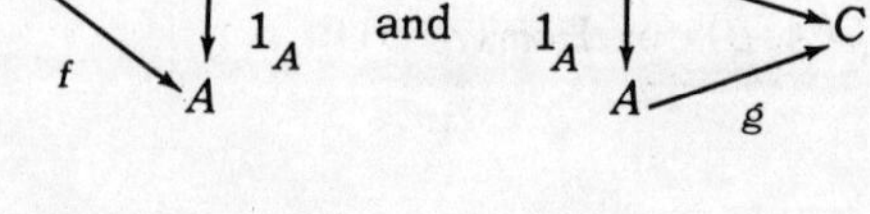

commuting for every $f \in \text{Hom}(B, A)$ and every $g \in \text{Hom}(A, C)$. This says $1_A f = f$, all f, and $g 1_A = g$, all g.

EXAMPLES.

1. $\mathcal{C}$ = all sets, all functions, ordinary composition of functions.
2. $\mathcal{C}$ = all groups, all homomorphisms, ordinary composition.
3. If R is a ring, $\mathcal{C}$ = all left R-modules, all R-homomorphisms, ordinary composition. This category is denoted ${}_R\mathfrak{M}$.
4. $\mathcal{C}$ = all topological spaces, all continuous maps, ordinary composition.
5. Let S be a semigroup with unit (i.e., a monoid). Define $\mathcal{C}$ as follows:

   obj $\mathcal{C} = \{A\}$ (i.e., only one object);

   $\text{Hom}(A, A) = S$, (i.e., the morphisms are just the elements of S);

   composition = given multiplication of S.

Note that $\text{Hom}(A, A)$ is not a set of functions.

At this stage, there is no need to drown yourself in examples. For the most part, we shall keep to the category of left R-modules. A good analogy to bear in mind is that of point-set topology. The definition of topological space is quite abstract; further, there are many natural examples, as well as many pathological, artificial examples. Next in line, one generally imposes restrictions on spaces (separation axioms, separability axioms, etc.) which take advantage of extra properties interesting spaces possess. This is, in fact, also done for categories. For example, in many reasonable categories, $\text{Hom}(A, B)$ has more structure than merely being a set, e.g., it is often an abelian group. Further, one might want to have various useful constructions permissible, e.g., direct products. Finally, the *raison d'etre* of topological spaces is that continuous functions live on them; the *raison d'etre* of categories is that functors live on them. If one thinks of categories as generalized monoids ("generalized" because multiplication of elements is not always defined), then a functor is just a homomorphism.

DEFINITION. Let $\mathcal{C}$ and $\mathcal{C}'$ be categories. A **functor** $F: \mathcal{C} \to \mathcal{C}'$ is a function satisfying:

1) If $A \in \text{obj}\ \mathcal{C}$, then $FA \in \text{obj}\ \mathcal{C}'$;
2) If $f: A \to B$ in $\mathcal{C}$, then $Ff: FA \to FB$ in $\mathcal{C}'$;
3) $A \xrightarrow{f} B \xrightarrow{g} C$ in $\mathcal{C}$, then $F(gf) = Fg \cdot Ff$;
4) $F(1_A) = 1_{FA}$.

Before giving examples, here are some elementary properties.

THEOREM 1.1. *For* $A \in \text{obj}\ \mathcal{C}$, *there is a unique* $1_A \in \text{Hom}(A, A)$.

Proof: Suppose 2_A is a second identity morphism. Then, since 1_A is an identity, $1_A = 1_A 2_A$; also $1_A 2_A = 2_A$

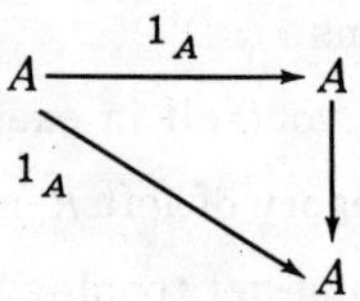

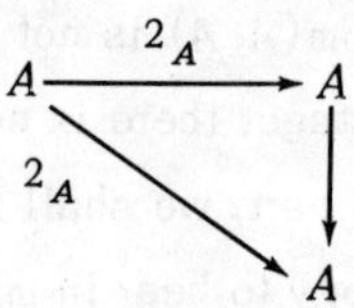

because 2_A is an identity. Hence $1_A = 2_A$.

The function $A \mapsto 1_A$ is thus a one-to-one correspondence between objects and identity morphisms. One could, therefore, define a category using only two primitive data—morphisms and composition, dispensing entirely with objects. This is sometimes convenient, but usually it is not.

DEFINITION. A morphism $f: A \to B$ is an **isomorphism** if there is a morphism $g: B \to A$ with $gf = 1_A$ and $fg = 1_B$.

EXAMPLES. In groups, isomorphisms are isomorphisms. In topological spaces, isomorphisms are homeomorphisms.

THEOREM 1.2. *Let $F : \mathcal{C} \to \mathcal{C}'$ be a functor. If f is an isomorphism, so is Ff.*

Proof. If $fg = 1_B$ and $gf = 1_A$, then applying F gives $Ff \cdot Fg = 1_{FB}$ and $Fg \cdot Ff = 1_{FA}$, as desired.

We have defined a covariant functor. There are also contravariant functors (which reverse the direction of arrows).

DEFINITION. A contravariant functor $F : \mathcal{C} \to \mathcal{C}'$ is a function satisfying:

1) If $A \in \text{obj}\,\mathcal{C}$, then $FA \in \text{obj}\,\mathcal{C}'$;
2) If $f: A \to B$, then $Ff: FB \to FA$;
3) If $A \xrightarrow{f} B \xrightarrow{g} C$, then $F(gf) = Ff \cdot Fg: FC \to FA$;
4) $F(1_A) = 1_{FA}$.

EXERCISE 1.1. If $F : \mathcal{C} \to \mathcal{C}'$ is a contravariant functor, and if f is an isomorphism, then Ff is an isomorphism.

EXAMPLES 1. Define $U : \mathit{gps} \to$ sets by $U(G) =$ underlying set of G, and $Uf = f$ (think of a homomorphism as a mere function). U is called a **forgetful functor**.

2. Fix A. Then $F = \text{Hom}(A, \;\;): \mathcal{C} \to$ sets is a functor:

 i) $FB = \text{Hom}(A, B)$;

 ii) if $g: B \to C$, then $Fg: \text{Hom}(A, B) \to \text{Hom}(A, C)$ by $f \mapsto gf$.

3. Fix B. Then $F = \text{Hom}(\;\;, B): \mathcal{C} \to$ sets is a contravariant functor:

 i) $FA = \text{Hom}(A, B)$;

 ii) if $f: A \to A'$, then $Ff: \text{Hom}(A', B) \to \text{Hom}(A, B)$ by $h \mapsto hf$.

DEFINITION. A category $\mathcal{C}$ is **pre-additive** if

i) $\operatorname{Hom}(A, B)$ is an additive abelian group;

ii) the distributivity laws hold (when defined).

DEFINITION. A functor $F : \mathcal{C} \to \mathcal{C}'$, where $\mathcal{C}, \mathcal{C}'$ are pre-additive categories, is **additive** if $F(f+g) = Ff + Fg$.

Ordinarily, we tacitly assume F is additive when we say F is a functor between pre-additive categories.

Note that if ${}_R\mathfrak{M}$ = category of R-modules, then ${}_R\mathfrak{M}$ is pre-additive. Moreover, the Hom functors: ${}_R\mathfrak{M} \to$ sets are, in reality, functors: ${}_R\mathfrak{M} \to$ abelian groups (and they are additive functors).

EXAMPLE. Let A be a right R-module. Then $F : {}_R\mathfrak{M} \to$ abelian groups defined by $F(B) = A \otimes_R B$ and $Ff = 1_A \otimes f$, is an (additive) functor.

EXERCISE 1.2. In a pre-additive category, zero maps make sense. If F is an (additive) functor (contra or co), prove that $F(0) = 0$, where 0 is a zero morphism.

Hom and $\otimes$ are the most important functors in homological algebra. Before we study them more, we give a final construction of a functor. This example has nothing to do with our subject, but it is where the basic ideas originate. We construct the (singular) homology functors: topological spaces $\to$ abelian groups.

Let v_0 be the origin in real Hilbert space, and let $\{v_1, v_2, \ldots, v_n, \ldots\}$ be an orthonormal basis. For each $n \geq 0$, let Δ_n be the convex hull of the points $\{v_0, \ldots, v_n\}$. Thus, $\Delta_0 = [v_0]$ is a point, $\Delta_1 = [v_0, v_1]$ is a line with endpoints v_0, v_1; $\Delta_2 = [v_0, v_1, v_2]$ is a triangle; $\Delta_3 = [v_0, v_1, v_2, v_3]$ is a tetrahedron, and so forth.

If X is a topological space, an *n-simplex* is a continuous map $T : \Delta_n \to X$. For $n \geq 0$, the group of *n-chains*, denoted $S_n(X)$, is the free abelian group generated by all n-simplexes. In particular, $S_0(X)$ is just the free abelian group generated by X itself (for a 0-simplex is just a point of X). It is convenient to define $S_{-1}(X) = 0$.

Every n-simplex has a "boundary" consisting of $(n-1)$-simplexes or *faces*. First, consider the boundary of Δ_n (in the point-set topological sense). Boundary $\Delta_n = \cup_{i=0}^{n}[v_0, \ldots, \hat{v}_i, \ldots, v_n]$, where $\hat{}$ means "delete." For example, boundary Δ_2 consists of the edges $[v_0, v_1]$, $[v_1, v_2]$, and $[v_0, v_2]$.

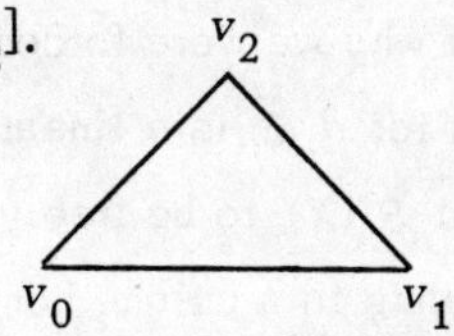

Observe that $[v_0, \ldots, \hat{v}_i, \ldots, v_n]$ is just the face opposite the vertex v_i.

An *orientation* of Δ_n is an ordering of its vertices, and we regard two orientations as the same if they have the same parity (both are even permutations or both are odd permutations).

An orientation of Δ_n induces an orientation of each of its faces in the following way: the ith face is oriented in the sense $(-1)^i[v_0, \ldots, \hat{v}_i, \ldots, v_n]$, where $-[v_0, \ldots, \hat{v}_i, \ldots, v_n]$ means this face with orientation opposite to that of $[v_0, \ldots, \hat{v}_i, \ldots, v_n]$. An example will show that this definition is quite natural. Consider Δ_2 oriented thus:

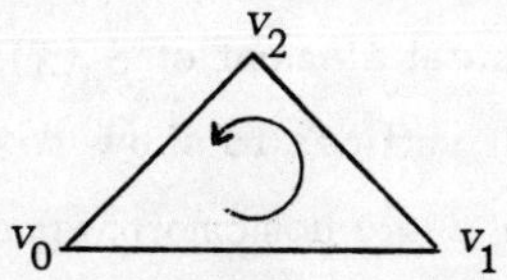

The natural way to orient the edges is thus:

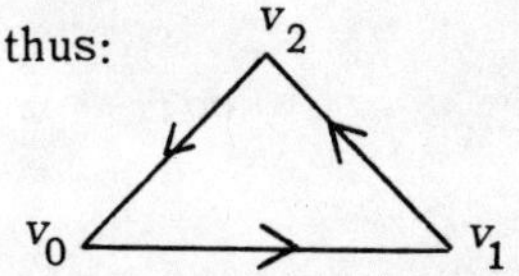

i.e., $[v_0, v_1]$, $[v_1, v_2]$, and $[v_2, v_0]$, and these are, respectively, the 2, 0 and 1 faces. But $[v_2, v_0] = -[v_0, v_2]$, so the "oriented boundary" of oriented Δ_2 is $[v_0, v_1]$, $-[v_0, v_2]$, and $[v_1, v_2]$.

Let us return to n-simplexes $T : \Delta_n \to X$. Define the ith face T^i of T to be $T|[v_0, \ldots, \hat{v}_i, \ldots, v_n]$. Better, let $e_i : \Delta_{n-1} \to \Delta_n$ be the map sending Δ_{n-1} into $[v_0, \ldots, \hat{v}_i, \ldots, v_n]$ and define $T^i = Te_i : \Delta_{n-1} \to X$, so that each T^i is an $(n-1)$-simplex. An obvious candidate for the boundary of T is

$$\partial_n T = \sum_{i=0}^{n} (-1)^i T^i \in S_{n-1}(X)$$

This is why we were forced to consider an abelian group of simplexes, for $\partial_n T$ is a linear combination of simplexes. Better yet, we want $S_n(X)$ to be freely generated by the simplexes, for then ∂_n extends to a unique homomorphism

$$\partial_n : S_n(X) \to S_{n-1}(X) .$$

We have arrived at the situation

$$\cdots \to S_n(X) \xrightarrow{\partial_n} S_{n-1}(X) \to \cdots \to S_1(X) \xrightarrow{\partial_1} S_0(X) \to S_{-1}(X) = 0 ,$$

where each ∂ is defined as an alternating sum of faces.

THEOREM 1.3. *For any n,* $\partial_{n-1}\partial_n = 0$.

Proof: Since a typical element of $S_n(X)$ is a linear combination of n-simplexes, it suffices to show $\partial\partial T = 0$ for each n-simplex T. Since the ∂ are homomorphisms,

$$\partial\partial T = \sum_{i=0}^{n} (-1)^i \partial T^i$$

$$= \sum_{i=0}^{n} (-1)^i \partial T[v_0, \ldots, \hat{v}_i, \ldots, v_n]$$

We choose to break the formula for ∂T^i into two pieces.

$$\partial\partial T = \sum_{i=0}^{n} (-1)^i \left\{ \sum_{j=0}^{i-1} (-1)^j T[v_0, \ldots, \hat{v}_j, \ldots, \hat{v}_i, \ldots, v_n] \right.$$

$$\left. + \sum_{j=i+1}^{n} (-1)^{j-1} T[v_0, \ldots, \hat{v}_i, \ldots, \hat{v}_j, \ldots, v_n] \right\}$$

(the exponent is $j-1$ in the second half instead of j because we have shifted the range of summation by 1). Observe that each $(n-2)$-dimensional face occurs as $[v_0, \ldots, v_n]$ with two vertices deleted; moreover, it occurs twice: if we call the deleted vertices a and b, the first occurrence is when $a = v_j$ and $b = v_i$; the second occurrence is when $a = v_i$ and $b = v_j$. In the first instance, the sign is $(-1)^{i+j}$; in the second instance, the sign is $(-1)^{i+j-1}$. Therefore, everything cancels in pairs, and $\partial\partial T = 0$.

Given a homomorphism one should (with Pavlov's dog) salivate: What is its kernel? What is its image? Let $Z_n(X) = \ker \partial_n$ (elements here are called *n-cycles*); let $B_n(X) = \operatorname{im} \partial_{n+1}$ (elements here are called *n-boundaries*). Both of these are subgroups of $S_n(X)$. Now $\partial_n \partial_{n+1} = 0$ implies that $B_n(X) \subset Z_n(X)$, so that the quotient $Z_n(X)/B_n(X)$ makes sense. We define $H_n(X)$, the nth *homology group* of X, by

$$H_n(X) = Z_n(X)/B_n(X) .$$

For any fixed $n \geq 0$, we claim that H_n is a functor from topological spaces to abelian groups. It remains to define, for any continuous $f: X \to Y$, a homomorphism $H_n(f): H_n(X) \to H_n(Y)$. This is done as follows. Define $f_\#: S_n(X) \to S_n(Y)$ by $T \mapsto f \circ T$. Now $\partial f_\# = f_\# \partial$, i.e., the following diagram commutes:

$$\begin{array}{ccc} S_n(X) & \xrightarrow{\partial_n} & S_{n-1}(X) \\ \downarrow f_\# & & \downarrow f_\# \\ S_n(Y) & \xrightarrow[\partial_n]{} & S_{n-1}(Y) \end{array}$$

It follows that $f_\#(Z_n(X)) \subset Z_n(Y)$ and $f_\#(B_n(X)) \subset B_n(Y)$, so that $f_\#$ induces a map on the quotients $H_n(X) \to H_n(Y)$, namely,

$$z_n + B_n(X) \mapsto f_\#(z_n) + B_n(Y) \; .$$

This map is, by definition, $H_n(f)$. We let the reader verify that each H_n is a functor. In particular, if $f: X \to Y$ is a homeomorphism, then $H_n(f)$ is an isomorphism; homeomorphic spaces have the same homology groups.

This construction is modified a bit by topologists. Suppose G is a fixed abelian group. Replace the sequence

$$\cdots \to S_n(X) \xrightarrow{\partial_n} S_{n-1}(X) \to \cdots$$

by the sequence

$$\cdots \to S_n(X) \otimes G \xrightarrow{\partial_n \otimes 1} S_{n-1}(X) \otimes G \to \cdots$$

Again the composite of any two successive maps is 0 $[(\partial_{n-1} \otimes 1)(\partial_n \otimes 1) = \partial_{n-1}\partial_n \otimes 1 = 0 \otimes 1 = 0]$, so that we may again define cycles, boundaries, and homology. The groups so obtained are denoted $H_n(X; G)$. In particular, our original construction yields the groups $H_n(X; Z)$.

A second useful modification constructs a contravariant functor (called *cohomology*). Here we replace the sequence

$$\cdots \to S_n(X) \xrightarrow{\partial_n} S_{n-1}(X) \to \cdots$$

by the sequence

$$\cdots \leftarrow \operatorname{Hom}(S_n(X), G) \xleftarrow{\operatorname{Hom}(\partial_n, G)} \operatorname{Hom}(S_{n-1}(X), G) \leftarrow \cdots$$

where G is a fixed abelian group. (Note that since Hom(, G) is contravariant, the arrows have changed direction.) Denote $\operatorname{Hom}(\partial_n, G)$ by δ_n (so that $\delta_n(f) = f \circ \partial_n$). Now $\delta_n \delta_{n-1} = 0$, so analogues of cycles, boundaries, and homology can be defined. We obtain contravariant functors, denoted H^n.

From time to time, it will help the reader to return to this section to see similarities.

Here is a glimpse of an algebraic context in which one constructs a long sequence of modules in which the composite of successive maps is 0. If M is an R-module, we know it can be described by generators and relations, i.e., there is an exact sequence $0 \to K_0 \to F_0 \to M \to 0$, where F_0 is a free module. Now K_0 need not be free, so describe it by generators and relations:

$$0 \to K_1 \to F_1 \to K_0 \to 0\ ,$$

where F_1 is a free module. Link these together to get

$$\begin{array}{ccccccccc} K_1 & \to & F_1 & & \to & & F_0 & \to M \to 0\ . \\ & & & \searrow & & \nearrow & & & \\ & & & & K_0 & & & & \end{array}$$

Now K_1 need not be free, so iterate this procedure. In general, this process can be continued indefinitely without arriving at a free kernel. In sum, we have an infinite sequence

$$\cdots \to F_n \to F_{n-1} \to \cdots \to F_0 \to M \to 0\ ,$$

where each F_n is a free module and the composite of successive maps is 0 (better, the sequence is exact). The analogy with topological spaces looks even better when we recall that $S_0(X)$ is just the free abelian group on X itself. This sequence will be exploited later.

§2. Hom *and* ⊗.

We shall concentrate on the category ${}_R\mathfrak{M}$ of left R-modules, where R is a fixed ring (always associative with 1), and we begin by considering the notions of sum and product. Since our main interest is the study of the functors Hom and ⊗, we shall try to get descriptions of sum and product that are amenable to such a study.

DEFINITION. Let $\{A_k : k \in K\}$ be a family of R-modules. The **product** of the A_k, denoted $\Pi_{k \in K} A_k$, is an R-module whose underlying set is the cartesian product of the A_k, i.e., all $a = (a_k)$, $a_k \in A_k$, with module operations

$$(a_k) + (b_k) = (a_k + b_k) ;$$
$$r(a_k) = (ra_k) .$$

DEFINITION. The **sum** of the A_k, denoted $\Sigma_{k \in K} A_k$ (or sometimes $\amalg_{k \in K} A_k$) is the submodule of ΠA_k consisting of all (a_k) almost all of whose coordinates are O (i.e., all but a finite number of a_k are O).

Note that if the index set K is finite, $\Pi_{k \in K} A_k = \Sigma_{k \in K} A_k$. If K is infinite, and the $A_k \neq O$, then ΣA_k is a proper submodule of ΠA_k. If K is finite, we also write $A_1 \oplus \cdots \oplus A_n$ for $\Pi_{k=1}^n A_k$.

Exercise 2.1. Give an example of a category of R-modules in which there is an infinite set of non zero modules $\{A_k : k \in K\}$ such that $\Sigma_{k \in K} A_k \cong \Pi_{k \in K} A_k$. (This is usually not the case.)

Given a product or a sum, there are certain injection and projection maps that can be defined: if $A = \Pi A_k$ or $A = \Sigma A_k$, let $p_k: A \to A_k$ be the projection onto the k-th coordinate, and let $i_k: A_k \to A$ put a_k into the k-th coordinate with O elseqhere. Note that $p_k i_k = 1_{A_k}$ and $p_j i_k = O$ if $j \neq k$.

THEOREM 2.1. *Let A and $\{A_k : k \in K\}$ be given. Then $A \cong \Sigma_{k \in K} A_k$ if and only if there are $i_k : A_k \to A$ such that, given any module X and any homomorphisms $f_k : A_k \to X$, there is a unique $\phi : A \to X$ with $\phi i_k = f_k$, all $k \in K$.*

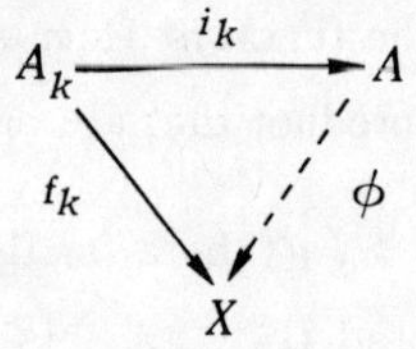

Proof: Suppose A is the sum of A_k, and let $\{p_k : A \to A_k\}$ be the projections. Define $\phi : A \to X$ by $\phi a = \Sigma f_k p_k a$. Note that since A is a sum, almost all $p_k a = O$, and so the formula for ϕa makes sense. The diagram does commute, for

$$\phi i_j a_j = \sum_k f_k p_k i_j a_j = f_j a_j .$$

We claim ϕ is the unique such map. Suppose $\psi : A \to X$ satisfies $\psi i_k = f_k$, all $k \in K$; if $a \in A$, $a = \Sigma i_k p_k a$, and so $\psi a = \Sigma \psi i_k p_k a = \Sigma f_k p_k a = \phi a$.

Suppose, conversely, that a unique $\phi : A \to X$ always exists. Consider the diagrams

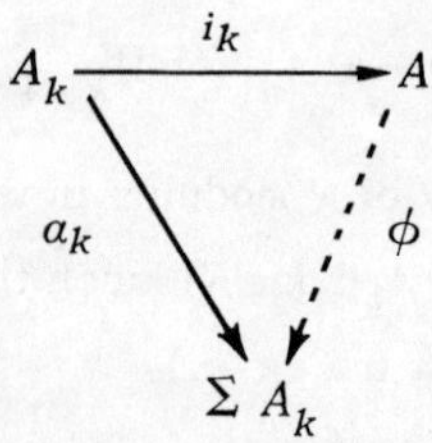

and

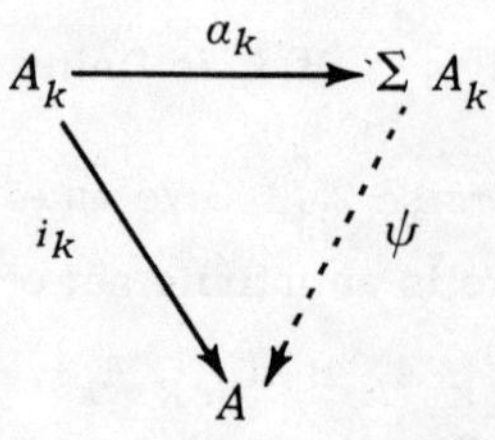

where $a_k : A_k \to \Sigma A_k$ is the injection. Now ϕ exists by hypothesis, while ψ exists by the first part of the proof. Assemble these diagrams into

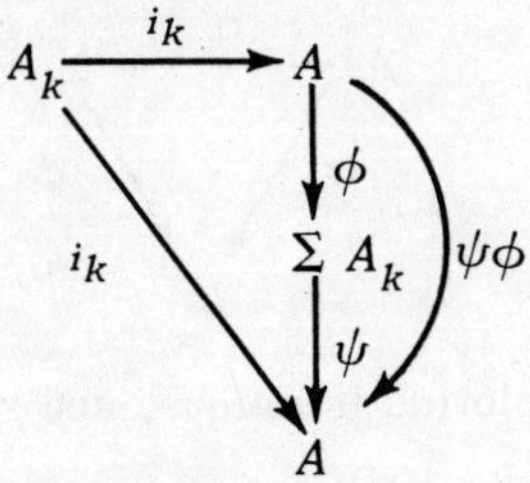

and note that we have commutativity: $\psi\phi i_k = \psi a_k = i_k$. This last diagram is one about which we have information ($X = A$ and $f_k = i_k$). In particular, we know that the map $\psi\phi : A \to A$ which completes it is unique. But, visibly, $1_A : A \to A$ completes the same diagram, so that $\psi\phi = 1_A$. If we assemble the original pair of diagrams a second way, then we obtain $\phi\psi = 1_{\Sigma A_k}$. Hence ϕ is an isomorphism.

REMARK. One can show that the maps i_k are all monic.

THEOREM 2.2. *Let* $\{A_k : k \in K\}$ *and* B *be modules, and let* $i_k : A_k \to \Sigma A_k$ *be injections. The map*

$$\text{Hom}(\Sigma A_k, B) \xrightarrow{\theta} \Pi \text{ Hom}(A_k, B)$$

given by

$$\phi \mapsto (\phi i_k)$$

is an isomorphism.

Proof: It is clear that θ is a homomorphism. θ is epic: if $(f_k) \in \Pi \text{ Hom}(A_k, B)$, then $f_k : A_k \to B$, all $k \in K$. By Theorem 2.1, there is a map $\phi : \Sigma A_k \to B$ with $\phi i_k = f_k$, all $k \in K$, i.e., $\theta(\phi) = (f_k)$. θ is monic: suppose $\theta(\phi) = (\phi i_k) = O$, i.e., each $\phi i_k = O$. Thus ϕ completes the diagram

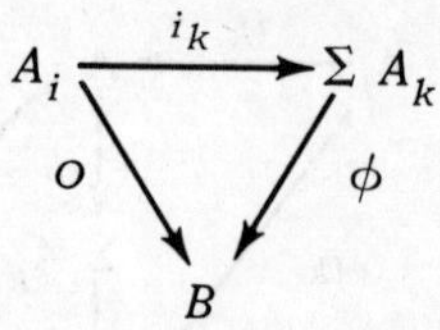

Since any such completion is unique, and since the zero map also completes, we have $\phi = O$.

If we denote the functor Hom(, B) by F, then we see that $F(\Sigma A_k) \cong \Pi F(A_k)$, i.e., F converts sums to products. In particular, F preserves *finite* sums.

Exercise 2.2. Give an example of $F = \text{Hom}(\ \ , B)$ for which $F(\Pi A_k) \not\cong \Pi F(A_k)$.

Let us now give the same treatment to products.

THEOREM 2.3. *Let A and $\{A_k : k \in K\}$ be given. Then $A \cong \Pi A_k$ if and only if there are $p_k : A \to A_k$ such that, given any module X and any homomorphisms $f_k : X \to A_k$, there is a unique $\phi : X \to A$ with $p_k \phi = f_k$, all $k \in K$.*

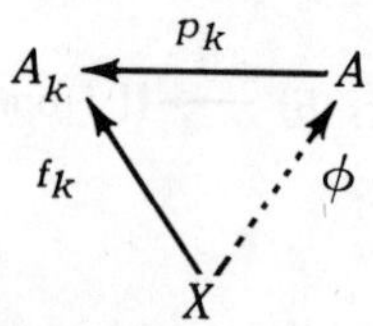

Proof: Suppose A is the product of the A_k, and let $i_k : A_k \to A$ be the injections. Define $\phi : X \to A$ by $\phi x = (f_k x)$. The diagram does commute, for $p_k \phi x = p_k(f_k x) = f_k x$. Such a ϕ is unique, for if $\psi : X \to A$ satisfies $p_k \psi = f_k$, then $p_k \psi x = f_k x$ and so $\psi x = (f_k x) = \phi x$.

Suppose, conversely, that a unique $\phi : X \to A$ always exists. Consider the diagrams

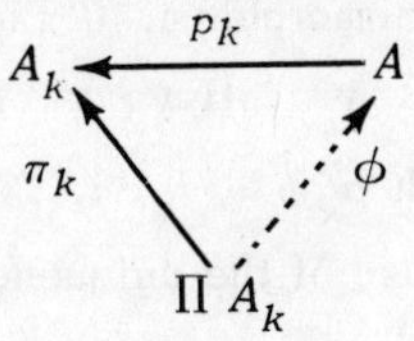

and

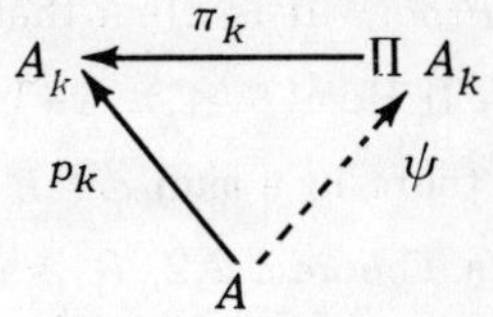

where $\pi_k : \Pi A_k \to A_k$ is the projection. Now ϕ exists by hypothesis, while ψ exists by the first part of the proof. Assemble these diagrams into

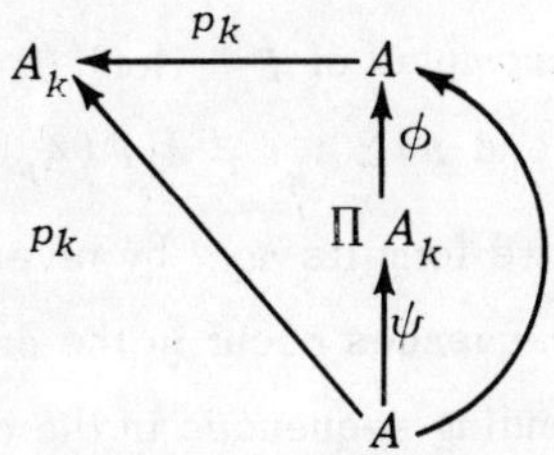

and note that we have commutativity: $p_k \phi\psi = \pi_k \psi = p_k$. This last diagram is one about which we have information ($X = A$ and $f_k = p_k$). In particular, we know that the map $\phi\psi : A \to A$ which completes the diagram is unique. But, visibly, $1_A : A \to A$ completes the same diagram, so that $\phi\psi = 1_A$. Similarly, $\psi\phi = 1_{\Pi A_k}$, and so ϕ is an isomorphism.

REMARK: One can show that the maps p_k are epic.

I hope the last proof induced a sense of déja vu in the reader. More about that in a minute.

THEOREM 2.4. *Let* $\{A_k : k \in K\}$ *and* B *be modules, and let* $p_k : \Pi A_k \to A_k$ *be projections. The map*

$$\operatorname{Hom}(B, \Pi A_k) \xrightarrow{\theta} \Pi \operatorname{Hom}(B, A_k)$$

given by

$$\phi \mapsto (p_k \phi)$$

is an isomorphism.

Proof: It is clear that θ is a homomorphism. θ is epic: given $(f_k) \in \Pi \operatorname{Hom}(B, A_k)$, we have $f_k : B \to A_k$, all $k \in K$. By Theorem 2.3, there is a map $\phi: B \to \Pi A_k$ with $p_k \phi = f_k$, i.e., $\theta(\phi) = (f_k)$. As in Theorem 2.2, θ is monic because of the uniqueness of ϕ.

If we denote the functor Hom$(B,\ \ )$ by F, we have just seen that $F(\Pi A_k) \cong \Pi F(A_k)$, i.e., F preserves products. In particular, F preserves *finite* sums.

Exercise 2.3. Give examples of $F = \operatorname{Hom}(B,\ \ )$ for which $F(\Sigma A_k) \not\cong \Sigma F(A_k)$, and $F(\Sigma A_k) \not\cong \Pi F(A_k)$.

Given a diagram, we form its *dual* by reversing the direction of all arrows; if exact sequences occur in the original diagram, we insist that the corresponding sequences in the dual diagram be exact; finally, if any portion of the original diagram commutes, we insist that the corresponding portion of the dual diagram also commutes.

EXAMPLES: "$A \to B \to O$ exact" has as its dual diagram "$A \leftarrow B \leftarrow O$ exact" or, "$O \to B \to A$ exact." In words, the dual of epic is monic. Similarly, the dual of monic is epic. A comparison of Theorems 2.1 and 2.3 shows that sum and product are dual. We shall see more examples soon.

It is not true that the dual of a theorem is also a theorem (where one dualizes words as we have done in the examples). However, should both be true, it is often the case that the proofs of each are the same, better, that they are dual. A case in point is the pair of theorems just proved for sums and for products. In future instances, we will only supply at most one of the proofs.

The special case of sum of two modules merits more discussion.

THEOREM 2.5. *Given A and B with $i: A \to B$ monic, then A is a summand of B (i.e., $B = iA \oplus C$ for some submodule C of B) if and only if there is a map $p: B \to A$ with $pi = 1_A$.*

Proof: If A is a summand, let $p: B \to A$ be the projection on A (having kernel C). Conversely, if a map $p: B \to A$ exists, define $C = \ker p$. We claim $B = iA \oplus C$. If $b \in B$, then

$$b = ipb + (b - ipb).$$

The first term is in iA, the second in C, and clearly $iA \cap C = O$.

DEFINITION. A short exact sequence

$$O \longrightarrow A \xrightarrow{i} B \xrightarrow{p} C \longrightarrow O$$

is **split** if there is a map $j: C \to B$ with $pj = 1_C$.

Exercises. 2.4. The sequence above is split if and only if there is a map $q: B \to A$ with $qi = 1_A$.

2.5. Suppose $A = B \oplus C$ and $B \subset M \subset A$. Then $M = B \oplus (M \cap C)$.

Note that a short exact sequence is split if and only if A is a summand of B with the given map $i: A \to B$ as the injection map. In particular, $B \cong A \oplus C$.

A consequence of Theorems 2.2 and 2.4 is that either functor $\operatorname{Hom}(A, \ )$ or $\operatorname{Hom}(\ , B)$ takes split short exact sequences into split short exact sequences.

We assume the reader recalls how tensor product behaves with respect to sums: $A \otimes \Sigma B_k \cong \Sigma(A \otimes B_k)$ and $(\Sigma A_i) \otimes B \cong \Sigma(A_i \otimes B)$. It is important to know that the isomorphisms are $a \otimes \Sigma b_k \to \Sigma(a \otimes b_k)$ and $(\Sigma a_i) \otimes b \mapsto \Sigma(a_i \otimes b)$ respectively. Thus, for example, if $F = A\otimes$, then $F(\Sigma B_k) \cong \Sigma F(B_k)$ and F preserves sums; in particular, F preserves split short exact sequences.

Exercises 2.6. Give an example of $F = A\otimes$ for which

$$F(\Pi B_k) \not\cong \Pi F(B_k).$$

2.7. If F is any additive functor (contra- or co-), then

$$F\left(\sum_{i=1}^{n} A_i\right) \cong \sum_{i=1}^{n} F(A_i).$$

In particular, any additive functor preserves split short exact sequences.

Note: If we neglect specifying a functor's domain and target, it is tacitly assumed they are the categories of R-modules and S-modules respectively, where R and S are rings.

DEFINITION. A functor F is **left exact** if exactness of

$$(*) \qquad O \to A \xrightarrow{\alpha} B \xrightarrow{\beta} C$$

implies exactness of

$$O \to FA \xrightarrow{F\alpha} FB \xrightarrow{F\beta} FC;$$

a functor F is **right exact** if exactness of

$$A \to B \to C \to O$$

implies exactness of

$$FA \to FB \to FC \to O.$$

Exercises 2.8. A functor F is left exact if and only if F "preserves kernels," i.e., for any morphism β, $F(\ker\beta) \cong \ker F\beta$, where, in the notation of sequence (*), the isomorphism is $F\alpha$. In particular, F preserves monomorphisms.

2.9. A functor F is right exact if and only if, for any morphism α, $F(\text{coker } \alpha) = \text{coker } F\alpha$ (as in the preceding exercise, interpret the equality as a suitable isomorphism). In particular, F preserves epimorphisms.

There are analogous definitions for contravariant functors.

DEFINITION. A contravariant functor F is **left exact** if exactness of

$$A \to B \to C \to O$$

implies exactness of

$$O \to FC \to FB \to FA\,;$$

a contravariant functor is **right exact** if exactness of

$$O \to A \to B \to C$$

implies exactness of

$$FC \to FB \to FA \to O\,.$$

DEFINITION. A functor is **exact** if it is both left exact and right exact.

Exercise 2.10. Let F be contravariant. F is left exact if and only if, for any morphism f, $F(\text{coker } f) = \ker Ff$; F is right exact if and only if, for any morphism f, $F(\ker f) = \text{coker}(Ff)$. As in the preceding two exercises, equality is not to be taken literally.

THEOREM 2.6. Hom$(M,\ \ )$ *is a left exact functor and* Hom$(\ \ ,N)$ *is a contravariant left exact functor.*

Proof: We shall only prove that $F = \text{Hom}(M, \quad)$ is left exact; the other half of the proof for $\text{Hom}(\quad, N)$ is similar.

Let $O \to A \xrightarrow{\alpha} B \xrightarrow{\beta} C$ be exact; we must show exactness of

$$O \longrightarrow \text{Hom}(M, A) \xrightarrow{F\alpha} \text{Hom}(M, B) \xrightarrow{F\beta} \text{Hom}(M, C),$$

where $(F\alpha)\phi = \alpha\phi$ and $(F\beta)\psi = \beta\psi$.

a) $F\alpha$ is monic.

If $\alpha\phi = O$, then $\alpha(\phi a) = O$, all $a \in A$. Since α is monic, $\phi a = O$, all $a \in A$, and so $\phi = O$.

b) $\text{im}\, F\alpha \subset \ker F\beta$.

Suppose $\psi \in \text{im}\, F\alpha$; then $\psi = \alpha\phi$, for some $\phi \in \text{Hom}(M, A)$. But then $(F\beta)\psi = \beta\psi = \beta\alpha\phi = O$ since $\beta\alpha = O$.

c) $\ker F\beta \subset \text{im}\, F\alpha$.

Suppose $\psi: M \to B$ and $\beta\psi = O$; we must show $\psi = \alpha\phi$ for some $\phi: M \to A$. We claim that if $m \in M$, then $\psi m \in \alpha A$, for $\beta(\psi m) = O$ and so $\psi m \in \ker\beta = \text{im}\,\alpha$. Hence $\alpha^{-1}\psi(m)$ makes sense and lies in A. Define $\phi: M \to A$ by $\phi(m) = \alpha^{-1}\psi(m)$, and observe that $\alpha\phi = \psi$.

THEOREM 2.7. *The functors $M\otimes$ and $\otimes N$ are right exact functors.*

Proof: We only give the proof for $F = M\otimes$; the other proof is similar. If $A \xrightarrow{\alpha} B \xrightarrow{\beta} C \to O$ is exact, we must show that

$$M \otimes A \xrightarrow{1\otimes\alpha} M \otimes B \xrightarrow{1\otimes\beta} M \otimes C \longrightarrow O$$

is exact.

a) $\text{im}\ 1\otimes\alpha \subset \ker\ 1\otimes\beta$.

It suffices to prove $(1\otimes\beta)(1\otimes\alpha) = O$. But

$$(1\otimes\beta)(1\otimes\alpha) = 1\otimes\beta\alpha = 1\otimes O = O.$$

b) $\ker(1\otimes\beta) \subset \operatorname{im}(1\otimes\alpha)$.

If $E = \operatorname{im}(1\otimes\alpha)$, then β induces a map $\bar{\beta} : (M\otimes B)/E \to M \otimes C$ by $m \otimes b + E \mapsto m \otimes \beta b$ (for $E \subset \ker(1\otimes\beta)$, by part (a)). Now if $\pi : M \otimes B \to (M\otimes B)/E$ is the natural map, it is easily seen that $\bar{\beta}\pi = 1 \otimes \beta$.

Suppose we show that $\bar{\beta}$ is an isomorphism. Then

$$\ker(1\otimes\beta) = \ker \bar{\beta}\pi = \ker \pi = E = \operatorname{im}(1\otimes\alpha) ,$$

and we are done. To show $\bar{\beta}$ is an isomorphism, we shall construct a map $M \otimes C \to (M \otimes B)/E$ inverse to it. The function

$$f: M \times C \to (M \otimes B)/E$$

given by

$$(m, c) \mapsto m \otimes b + E, \quad \text{where } \beta b = c,$$

is well-defined: if $\beta b' = \beta b = c$, then $\beta(b' - b) = 0$, so that there is $a \in A$ with $\alpha a = b' - b$; hence $m \otimes b' - m \otimes b = m \otimes (b' - b) = (1 \otimes \alpha)(m \otimes a) \in E$. Since f is bilinear, the universal property of tensor product gives a homomorphism $\bar{f} : M \otimes C \to (M \otimes B)/E$ with $\bar{f}(m \otimes c) = m \otimes b + E$. Visibly $\bar{f}$ and $\bar{\beta}$ are inverse.

c) $1 \otimes \beta$ is epic.

Let $\Sigma\, m_i \otimes c_i \in M \otimes C$. Since β is epic, there are $b_i \in B$ with $\beta b_i = c_i$, all i. Hence $(1 \otimes \beta)(\Sigma\, m_i \otimes b_i) = \Sigma\, m_i \otimes \beta b_i = \Sigma\, m_i \otimes c_i$.

We pause to give examples showing that neither $\operatorname{Hom}(A, \quad)$ nor $\operatorname{Hom}(\quad , B)$ need be right exact, and that $A \otimes$ need not be left exact. In each case, let $R = Z$, so that we deal with abelian groups.

Consider the sequence

$$O \to Z \to Q \to Q/Z \to O,$$

where Q = rationals.

I. $\mathrm{Hom}(A,\ \ )$ need not be right exact.

Let A be cyclic of order 2. Then $\mathrm{Hom}(A, Q) \to \mathrm{Hom}(A, Q/Z)$ cannot be epic, for $\mathrm{Hom}(A, Q) = O$, while $\mathrm{Hom}(A, Q/Z) \neq O$.

II. $\mathrm{Hom}(\ \ , B)$ need not be right exact.

Let $B = Z$. Then $\mathrm{Hom}(Q, Z) \to \mathrm{Hom}(Z, Z)$ cannot be epic, for $\mathrm{Hom}(Q, Z) = O$ and $\mathrm{Hom}(Z, Z) \cong Z \neq O$.

III. $A \otimes$ need not be left exact.

Let A be any nonzero torsion group. Then $A \otimes Z \to A \otimes Q$ cannot be monic, for $A \otimes Z \cong A$ while $A \otimes Q = O$ (if $a \in A$, there is a nonzero integer n with $na = O$; hence, if $q \in Q$, we have

$$a \otimes q = a \otimes n(q/n) = na \otimes q/n = O \otimes q/n = O) .$$

There is a remarkable relationship between Hom and $\otimes$. A small preamble is needed first. If R and S are rings, we say A is a **2-sided module,** denoted ${}_SA_R$, if A is a left S-module, a right R-module, and there is an associative law

$$s(ar) = (sa)r \quad \text{for all } s \in S,\ a \in A,\ r \in R.$$

An example of such a situation is $A = R = S$, with actions ordinary multiplication. Given the situation ${}_SB_R$ and ${}_SC$ (a 2-sided module and a left module), then the group $\mathrm{Hom}_S(B, C)$ has more structure: it is a left R-module, where we define $(rf)b = f(br)$. Also, given the situation ${}_SB_R$ and ${}_RA$, the group $B \otimes_R A$ has more structure: it is a left S-module, where $s(b \otimes a) = (sb) \otimes a$.

THEOREM 2.8. *(Adjoint Isomorphism) Given the situation* ${}_R A$, ${}_S B_R$, *and* ${}_S C$, *there is an isomorphism*

$$\operatorname{Hom}_R(A, \operatorname{Hom}_S(B, C)) \cong \operatorname{Hom}_S(B \otimes_R A, C) .$$

Proof: Define ψ: left → right by

$$(\psi f)(b \otimes a) = f_a(b), \text{ where } f_a : B \to C ;$$

define ϕ: right → left by

$$(\phi g)_a(b) = g(b \otimes a)$$

We let the reader verify all is well-defined and that ϕ and ψ are inverse.

The reason this isomorphism is called "adjoint" is quite formal. Let $G = \operatorname{Hom}_S(B, \ \ )$ and let $F = B \otimes_R$. With this notation, Theorem 2.8 becomes $\operatorname{Hom}_R(A, GC) \cong \operatorname{Hom}_S(FA, C)$. If one pretends that $\operatorname{Hom}(\ \ ,\ \ )$ is an inner product, then G and F are, in the parlance of linear algebra, adjoint.

If $f: A' \to A$, we shall use the following notations. If $F = \operatorname{Hom}(X, \ \ )$, then write $Ff = f_*$; if $G = \operatorname{Hom}(\ \ , X)$, write $Gf = f^*$. As a mnemonic, lower star for covariant, upper star for contravariant.

DEFINITION. Let $F : \mathfrak{A} \to \mathfrak{B}$ and $G: \mathfrak{B} \to \mathfrak{A}$ be functors. The ordered pair G, F is an **adjoint pair** if, for each $A \in \mathfrak{A}$, $B \in \mathfrak{B}$, there is an isomorphism $\tau = \tau_{A,B}$

$$\operatorname{Hom}_{\mathfrak{A}}(A, GB) \to \operatorname{Hom}_{\mathfrak{B}}(FA, B)$$

which is natural in each variable, i.e., the following diagrams commute:

$$\begin{array}{ccc} \mathrm{Hom}_{\mathfrak{A}}(A, GB) & \xrightarrow{f^*} & \mathrm{Hom}_{\mathfrak{A}}(A', GB) \\ \downarrow \tau & & \downarrow \tau \\ \mathrm{Hom}_{\mathfrak{B}}(FA, B) & \xrightarrow{(Ff)^*} & \mathrm{Hom}_{\mathfrak{B}}(FA', B) \end{array} \quad \text{all } f: A' \to A \text{ in } \mathfrak{A} ;$$

$$\begin{array}{ccc} \mathrm{Hom}_{\mathfrak{A}}(A, GB) & \xrightarrow{(Gg)_*} & \mathrm{Hom}_{\mathfrak{A}}(A, GB'') \\ \downarrow \tau & & \downarrow \tau \\ \mathrm{Hom}_{\mathfrak{B}}(FA, B) & \xrightarrow{g_*} & \mathrm{Hom}_{\mathfrak{B}}(FA, B'') \end{array} \quad \text{all } g: B \to B'' \text{ in } \mathfrak{B} .$$

THEOREM 2.9. *If* $B = {}_S B_R$ *is a 2-sided module, then* $\mathrm{Hom}_S(B, \ )$ *and* $B \otimes_R$ *form an adjoint pair.*

Proof: We have already constructed the adjoint isomorphism in 2.8; it only remains to check naturality. This is not very exciting, and the straightforward computations are left to the reader.

It is amusing that adjointness and left exactness of Hom give a proof of the right exactness of tensor. Although the next two results hold for more general categories, we give the proofs only for categories of modules lest the reader be bored by many definitions.

Here is the converse of Theorem 2.6.

LEMMA 2.10. *Suppose*

$$(*) \qquad B' \xrightarrow{\alpha} B \xrightarrow{\beta} B'' \longrightarrow 0$$

is a sequence of homomorphisms (not necessarily exact!); suppose further that for each module M, *we have exactness of*

$$0 \longrightarrow \mathrm{Hom}(B'', M) \xrightarrow{\beta^*} \mathrm{Hom}(B, M) \xrightarrow{\alpha^*} \mathrm{Hom}(B', M) \ .$$

Then the original sequence (*) *is exact.*

REMARK: A dual statement holds for the second variable of Hom with the "O" in sequence (*) occurring at the left.

Proof: (i) β is epic: let $M = \operatorname{coker} \beta = B''/\operatorname{im} \beta$ and $f: B'' \to M$ be the natural map. Then $\beta^* f = f\beta = O$, so that $f = O$ since β^* is monic. Hence $\operatorname{coker} \beta = O$ and β is epic.

(ii) $\operatorname{im} \alpha \subset \ker \beta$: we know $\alpha^* \beta^* = O$. In particular, if $M = B''$ and $f = 1_{B''}$, then $\alpha^* \beta^* f = f\beta\alpha = O$. Hence $\beta\alpha = O$.

(iii) $\ker \beta \subset \operatorname{im} \alpha$: let $M = \operatorname{coker} \alpha = B/\operatorname{im} \alpha$ and let $g: B \to M$ be the natural map. Clearly $\alpha^* g = g\alpha = O$, so that exactness of the Hom sequence gives $f \in \operatorname{Hom}(B'', M)$ with $f\beta = g$. If $\operatorname{im} \alpha \neq \ker \beta$, there is an element $b \in B$ with $\beta b = O$ and $b \notin \operatorname{im} \alpha$; this second fact gives $gb \neq O$. But $gb = f\beta b = O$, a contradiction.

The intrusion of elements into the final portion of the proof is a lack of elegance, but we have already apologized for restricting our attention to modules.

THEOREM 2.11. *Let* $F: {}_R\mathfrak{M} \to {}_S\mathfrak{M}$ *and* $G: {}_S\mathfrak{M} \to {}_R\mathfrak{M}$ *be functors. If* G, F *form an adjoint pair, then* G *is left exact and* F *is right exact.*

Proof: Suppose $A' \to A \to A'' \to O$ is exact; is $FA' \to FA \to FA'' \to O$ exact? Consider the commutative (by the naturality condition) diagram

$$\begin{array}{ccccccc} O \to & \operatorname{Hom}_R(A'', GB) & \longrightarrow & \operatorname{Hom}_R(A, GB) & \longrightarrow & \operatorname{Hom}_R(A', GB) \\ & \downarrow \tau & & \downarrow \tau & & \downarrow \tau \\ O \to & \operatorname{Hom}_S(FA'', B) & \longrightarrow & \operatorname{Hom}_S(FA, B) & \longrightarrow & \operatorname{Hom}_S(FA', B) \end{array},$$

where B is any S-module. Since $\operatorname{Hom}_R(\ \ , GB)$ is contravariant left exact, the top row is exact, and so the bottom row is exact. Since B is arbitrary, the lemma gives exactness of $FA' \to FA \to FA'' \to O$.

The proof of the left exactness of G is similar, using the other naturality condition.

COROLLARY 2.12. *$B \otimes_R$ is right exact.*

Proof: We may consider the right R-module B as a 2-sided module ${}_Z B_R$, so that $B \otimes_R$ is adjoint to $\mathrm{Hom}_Z(B, \ )$.

Observe that we have a necessary condition that a functor have an adjoint—it must be left or right exact. This is not sufficient.

We recall the definition of inverse limit. Let $\{A_k : k \in K\}$ be a set of modules, and suppose that there are certain pairs of indices $(k, j) \in K \times K$ for which there are given maps $\phi_{kj} : A_k \to A_j$. Assume:

i) for all k, $\phi_{kk} : A_k \to A_k$ is 1_{A_k};
ii) if ϕ_{kj}, ϕ_{ji}, and ϕ_{ki} are all defined, then the diagram

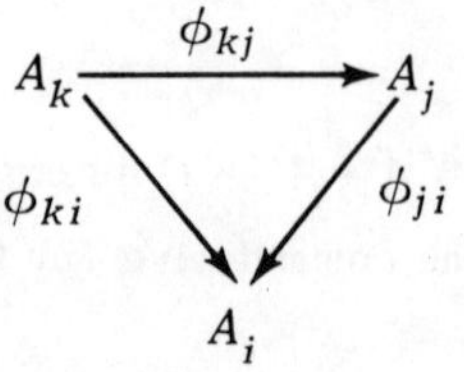

commutes.

DEFINITION. The **inverse limit**, $\varprojlim A_k$, is a module and a set of homomorphisms $\alpha_k : \varprojlim A_k \to A_k$ such that, for any module X, any $f_k : X \to A_k$ making the following diagram commute

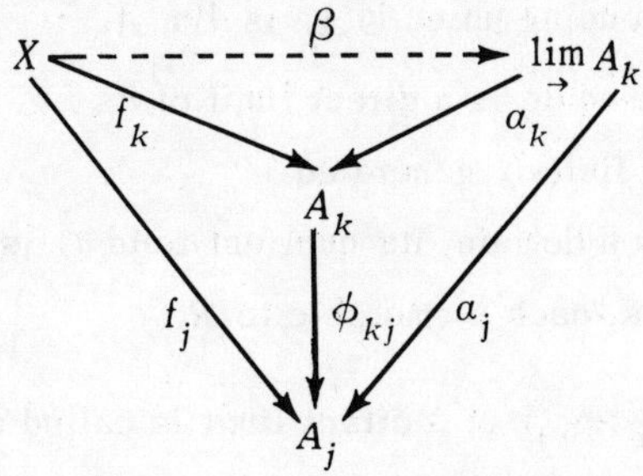

there is a unique $\beta: X \to \varprojlim A_k$ with all commuting.

Observe that if the set of certain pairs of indices is empty, $\varprojlim A_k = \Pi A_k$ (there are no maps ϕ_{kj} so that the lower triangles are not there, and we have the picture of Theorem 2.3). More generally, $\varprojlim A_k$ is the submodule of ΠA_k consisting of all (a_k) for which $\phi_{kj} a_k = a_j$ for each ϕ_{kj}; the maps α_k are just the projections.

There is a dual definition giving the **direct limit**, denoted $\varinjlim A_k$, whose formulation is left to the reader's mirror. Here, if the set of certain pairs of indices is empty, $\varinjlim A_k = \Sigma A_k$. More generally, $\varinjlim A_k$ is a quotient of ΣA_k, namely $(\Sigma A_k)/S$, where S is the submodule of ΣA_k generated by all $\phi_{kj} a_k - a_k$; the maps $\alpha_k: A_k \to \varinjlim A_k$ are just $a_k \mapsto a_k + S$.

A special instance of direct limit occurs when the index set K is **directed**, i.e., K is partially ordered and given $k, j \in K$, there is an element $i \in K$ with $k \leq i$ and $j \leq i$. In this case, we assume there are maps $\phi_{kj}: A_k \to A_j$ whenever $k \leq j$.

Exercises. In each of the following, show that the direct limits have directed index sets.

2.11 A sum $\Sigma_{k \in K} A_k$ is a direct limit of all finite "partial sums" $\Sigma_{i=1}^{n} A_{k_i}$.

2.12 An ascending union $\bigcup A_k$ is $\varinjlim A_k$.

2.13 Every module is a direct limit of its f.g. submodules. (f.g. abbreviates finitely generated.)

2.14 If R is a domain, its quotient field Q is a direct limit of cyclic modules, each isomorphic to R.

The data $\{A_k, \phi_{kj}\}$ of a direct limit is called a **direct system.** A **morphism** $\{A'_k, \phi'_{kj}\} \to \{A_k, \phi_{kj}\}$ between direct systems with the same index set is a family of maps $\{f_k : A'_k \to A_k, \text{ all } k\}$ making all the following diagrams commute:

$$\begin{array}{ccc} A'_k & \xrightarrow{f_k} & A_k \\ {\scriptstyle \phi'_{kj}}\downarrow & & \downarrow{\scriptstyle \phi_{kj}} \\ A'_j & \xrightarrow{f_j} & A_j \end{array} .$$

One checks easily that all direct systems (same index set!) and morphisms forms a category, Dir. We claim that $\varinjlim : \text{Dir} \to {}_R\mathfrak{M}$ is a functor: its value on an object $\{A_k, \phi_{kj}\}$ is $\varinjlim A_k = (\Sigma A_k)/S$; its value on a morphism $\{f_k : A'_k \to A_k\}$ is the map $\Sigma a'_k + S' \mapsto \Sigma f_k a'_k + S$.

The reason for introducing directed index sets is the following theorem, which says that $\varinjlim$ is an exact functor.

THEOREM 2.13 *Suppose there are morphisms of direct systems (with the same directed index set K)*

$$\{A'_k, \phi'_{kj}\} \xrightarrow{f} \{A_k, \phi_{kj}\} \xrightarrow{g} \{A''_k, \phi''_{kj}\}$$

such that, for each $k \in K$,

$$O \longrightarrow A_k' \xrightarrow{f_k} A_k \xrightarrow{g_k} A_k'' \longrightarrow O$$

is exact. Then we have exactness of

$$O \longrightarrow \varinjlim A_k' \xrightarrow{f} \varinjlim A_k \xrightarrow{g} \varinjlim A_k'' \longrightarrow O \ .$$

Proof: The only difficult part of the proof, and the only part requiring that the index set K be directed, is showing f is monic. We prove this and leave the remainder to the reader.

First, if $x \in \varinjlim A_k$, we claim there is an index i and some $a_i \in A_i$ with $x = a_i + S$. We know $x = y + S$, where $y \in \Sigma A_k$; thus $y = \Sigma_{j=1}^n a_{k_j}$. Since K is directed, there is an index $i \geq k_j$, all j. Define $a_j = \phi_{k_j i} a_{k_j} \in A_i$, and let $y' = \Sigma_{j=1}^n a_j \in A_i$. Since $a_j - a_{k_j} \in S$, we have $y' - y \in S$, and so $x = y + S = y' + S$.

Second, we identify the zero element of $\varinjlim A_k$: if $a_k \in A_k$ and $a_k + S = O$ in $\varinjlim A_k$, then there is some index $t \geq k$ with $\phi_{kt} a_k = O$. To see this, observe that elements of S have the form $\Sigma (\phi_{ji} a_j - a_j)$ (for any multiple of $\phi_{ji} a_j - a_j$ has the same form). Let t be an index with $t \geq k$, $t \geq$ each i, j occurring in this sum. Then

$$\phi_{kt} a_k = (\phi_{kt} a_k - a_k) + a_k = (\phi_{kt} a_k - a_k) + \Sigma (\phi_{ji} a_j - a_j) \, .$$

We can rewrite each of the terms on the right as a sum of relators in which the second index is t:

$$\phi_{ji} a_j - a_j = (\phi_{jt} a_j - a_j) + [\phi_{it}(-\phi_{ji} a_j) - (-\phi_{ji} a_j)] \ .$$

We now have

$$\phi_{kt} a_k = \Sigma (\phi_{jt} a_j - a_j) \ .$$

Now combine all terms having the same first index j (getting an element of the same form), so that, finally,

$$\phi_{kt} a_k = \sum_j (\phi_{jt} a_j - a_j) .$$

This is an equation in the sum ΣA_k (where every element has a unique expression as a sum of elements from each A_k). It follows that if $j \neq t$, then $a_j = O$. Since $\phi_{tt} a_t - a_t = O$, every term on the right side is O, hence $\phi_{kt} a_k = O$.

Let us now show that f is monic. If $fx = O$ in $\varinjlim A_k$, then first, $fx = f_k a_k' + S$ for some k, some $a_k' \in A_k'$. Second $f_k a_k' + S = O$ in $\varinjlim A_k$ implies $\phi_{kt} f_k a_k' = O$ for some t. But $\phi_{kt} f_k a_k' = f_t \phi_{kt}' a_k'$. Since f_t is monic, $\phi_{kt}' a_k' = O$, i.e., $a_k' + S' = O$ in $\varinjlim A_k'$. Hence $x = a_k' + S' = O$ and f is monic.

We shall have occasion to use two facts proved above:

i) each element of $\varinjlim A_k$ has a representative of the form $a_k + S$;

ii) an element $a_k + S$ is O if and only if some $\phi_{kj} a_k = O$.

THEOREM 2.14. *If F is a left exact functor that preserves products, then F preserves inverse limits.*

Proof: Define, for each special pair of indices (k, j), a module $A_{(k,j)}$ isomorphic to A_k. Define

$$f\colon \prod_k A_k \to \prod_{(k,j)} A_{(k,j)}$$

by

$$a \mapsto (p_k a - \phi_{jk} p_j a)$$

where p_k, p_j, are projections of the product $\Pi_k A_k$. Observe that

$\ker f = \varprojlim A_k$. Since F is left exact, F preserves kernels, i.e., $F(\varprojlim A_k) = \ker(Ff)$ (the equality is really an isomorphism, as in *Exercise* 2.8). Since F preserves products,

$$Ff\colon \prod FA_k \to \prod FA_{(k,j)}$$

and, if $\pi_{(k,j)}$ is a projection of the second product, then

$$\pi_{(k,j)} f = p_k - \phi_{jk} p_j$$

and

$$F\pi_{(k,j)} Ff = Fp_k - F\phi_{jk} Fp_j\,.$$

It follows that $\ker Ff = \varprojlim FA_k$, as desired.

COROLLARY 2.15. *For any A,* Hom$(A, \quad)$ *preserves inverse limits.*

Proof: We know Hom$(A, \quad)$ is left exact and preserves products.

The dual proof works for direct limits.

THEOREM 2.16. *If F is a right exact functor preserving sums, then F preserves direct limits.*

COROLLARY 2.17. *For any A, $A\otimes$ preserves direct limits.*

Exercises 2.15. The map $t_A\colon \mathrm{Hom}_R(R, A) \to A$ defined by $f \mapsto f(1)$ is an R-isomorphism.

2.16. Given $\phi\colon A \to B$, the following diagram commutes:

$$\begin{array}{ccc} A & \xrightarrow{\ \phi\ } & B \\ \uparrow{\scriptstyle t_A} & & \uparrow{\scriptstyle t_B} \\ \mathrm{Hom}_R(R, A) & \xrightarrow{\ \phi_*\ } & \mathrm{Hom}_R(R, B) \end{array}$$

2.17. The map $s_A : R \otimes_R A \to A$ defined by $r \otimes a \mapsto ra$ is an R-isomorphism.

2.18. Given $\phi : A \to B$, the following diagram commutes;

$$\begin{array}{ccc} A & \xrightarrow{\phi} & B \\ \uparrow{\scriptstyle s_A} & & \uparrow{\scriptstyle s_B} \\ R \otimes_R A & \xrightarrow{1 \otimes \phi} & R \otimes_R B \end{array}$$

2.19. Let $\{B_k, \phi_{kj}\}$ be a direct system. Prove that the map

$$A \otimes \varinjlim B_k \to \varinjlim (A \otimes B_k)$$

given by

$$a \otimes (\Sigma\, b_k + S) \mapsto \Sigma\, a \otimes b_k + S_o\,,$$

(where S, S_o are the relations in the construction of direct limit) is an isomorphism.

We mention a result of Watts (*Proc. AMS*, 1960) characterizing the functors Hom and $\otimes$.

(1) Let $F : {}_R\mathfrak{M} \to {}_Z\mathfrak{M}$ be an additive, covariant functor. If F is right exact and preserves sums, then $F = A \otimes_R$ for some R-module A.

(2) Let $F : {}_R\mathfrak{M} \to {}_Z\mathfrak{M}$ be an additive covariant functor. If F is left exact and preserves products, then $F = \text{Hom}_R(A, \quad)$ for some R-module A.

(3) Let $F : {}_R\mathfrak{M} \to {}_Z\mathfrak{M}$ be an additive, contravariant functor. If F is left exact and converts sums to products, then $F =$ $\text{Hom}_R(\quad, A)$ for some R-module A.

§3. *Projectives, Injectives, and Flats.*

DEFINITION. An R-module A is **free** if it is a sum of copies of R. If $Ra_k \cong R$ and $A = \Sigma_{k \epsilon K} Ra_k$, then the set $\{a_k: k \epsilon K\}$ is called a **basis** of A (or a *free set of generators* of A).

If $\{a_k: k \epsilon K\}$ is a basis of a free module A, then for each $a \epsilon A$, $a = \Sigma_k r_k a_k$, where $r_k \epsilon R$ and almost all $r_k = 0$. Moreover, there is only one such expression for any $a \epsilon A$.

THEOREM 3.1. *Let* $X = \{a_k: k \epsilon K\}$ *be a basis of a free module* A. *Given any module* B *and any function* $\alpha: X \to B$, *there is a unique homomorphism* $f: A \to B$ *extending* α.

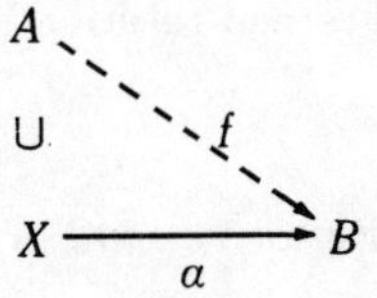

Proof: For fixed k, α defines a homomorphism $f_k: Ra_k \to B$ by $ra_k \mapsto r\alpha(a_k)$. Since A is the sum of $\{Ra_k: k \epsilon K\}$, Theorem 2.1 gives a unique $f: A \to B$ with $fa_k = f_k a_k = \alpha a_k$, all k.

Note that it is easy to describe f explicitly: a typical element of A is $\Sigma\, r_k a_k$, and f sends this element into $\Sigma\, r_k \alpha(a_k)$.

THEOREM 3.2. *If* X *is a set, then there is a free module* F *having* X *as a basis.*

Proof: If $X = \phi$, set $F = 0$. If $X \neq \phi$, let F be the sum of $|X|$ copies of R (where $|X|$ denotes the cardinal of X). There is a one-to-one correspondence between a basis of F and X; some relabeling lets us assume the basis is actually X.

THEOREM 3.3. *Every module A is a quotient of a free module.*

Proof: Let UA be the underlying set of A, and let F be the free module with basis UA. Let $\alpha: UA \to A$ be the "identity," i.e., $\alpha a = a$. By Theorem 3.1, there is a homomorphism $f: F \to A$ extending α, and f is epic because α is onto.

In down-to-earth language, Theorem 3.3 says A can be described by generators and relations: the generators are a basis of F, the relations comprise $\ker f$. Unfortunately, $\ker f$ need not be free, so that knowledge of a set of generators of $\ker f$ (even a minimal such set) need not describe $\ker f$ up to isomorphism.

Before pursuing generators and relations further, let us dispose of a quite natural question.

THEOREM 3.4. *If R is commutative, and A is a free R-module, then any two bases of A have the same cardinal.*

Proof: By Zorn's Lemma, R has a maximal ideal M. If $\{a_k: k \in K\}$ is a basis of A, let B_k denote $Ra_k/MRa_k \cong R/M$; denote the image of a_k in B_k by $\bar{a}_k$. We claim $A/MA \cong \Sigma_{k\in K} B_k$, for this follows from Theorem 2.1 if we define $i_k: B_k \to A/MA$ by $\bar{a}_k \mapsto a_k + MA$. But A/MA is an R-module annihilated by M, i.e., an R/M-module. Since M is maximal, R/M is a field, and so A/MA is a vector space over R/M. The isomorphism $A/MA \cong \Sigma_{k\in K} B_k$ shows that dimension $A/MA = |K|$, and so any two bases of A have cardinal $|K|$.

There are noncommutative rings for which Theorem 3.4 holds, but there do exist rings (Leavitt, *Proc. AMS,* 1956) for which the theorem is false. In fact, there is a ring R with $R \cong R \oplus R$ (the isomorphism, of course, considers both sides only as R-modules).

Exercise 3.1. Let R be commutative and let S be an *augmented* R*-algebra,* i.e., an R-algebra admitting a nonzero R-algebra map $\varepsilon: S \to R$. (Example: a group algebra RG.) Prove that Theorem 3.4 holds for free S-modules.

DEFINITION. A **free resolution** of a module A is an exact sequence $\cdots \to F_n \xrightarrow{d_n} F_{n-1} \to \cdots \cdots \to F_1 \xrightarrow{d_1} F_0 \to A \to O$ in which each F_n is a free module.

Note that in an exact sequence the composite of any two successive maps is O, i.e., im d_{n+1} = ker d_n implies im $d_{n+1} \subset$ ker d_n.

THEOREM 3.5. *Every module A has a free resolution.*

Proof: There is a free module F_0 and an exact sequence

$$O \longrightarrow S_o \longrightarrow F_o \xrightarrow{\varepsilon} A \longrightarrow O,$$

by Theorem 3.3. Similarly, there is a free module F_1 with exactness of

$$O \longrightarrow S_1 \longrightarrow F_1 \longrightarrow S_0 \longrightarrow O,$$

and, by induction, a free module F_n with exactness of

$$O \longrightarrow S_n \longrightarrow F_n \longrightarrow S_{n-1} \longrightarrow O.$$

Assemble all of these sequences into the diagram

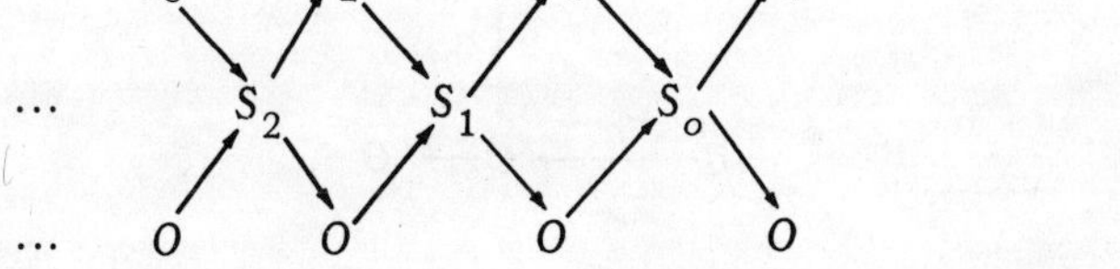

where the maps d_n are just the indicated composites. For any n, $\ker d_n = S_n$ and $\operatorname{im} d_n = S_{n-1}$. In particular, $\ker d_n = \operatorname{im} d_{n+1}$, and so the top row is exact.

Several comments are in order. First of all, a given module A has many free resolutions. Second, suppose we imitate the construction of homology groups: let n-cycles = $\ker d_n$, n-boundaries = $\operatorname{im} d_{n+1}$, and define homology as $\ker d_n / \operatorname{im} d_{n+1}$. All makes sense, but nothing happens: since a free resolution is exact, the quotient groups are all O.

DEFINITION. A **complex** is a sequence of modules and homomorphisms

$$\cdots \longrightarrow S_n \xrightarrow{d_n} S_{n-1} \xrightarrow{d_{n-1}} S_{n-2} \longrightarrow \cdots .$$

(indexed by all integers n, possibly negative) such that $d_{n-1} d_n = O$.

A free resolution is a complex, and its homology is O. In fact, the homology groups "measure" the deviation from exactness. Incidentally, it will be convenient to have the sequence be infinitely long to the right and to the left; we may think of a free resolution of A as doubly infinite, but with all modules to the right of A being O. Later on, we will spend more time investigating homology. We now return to free modules.

THEOREM 3.6. *Consider the following diagram with exact row*

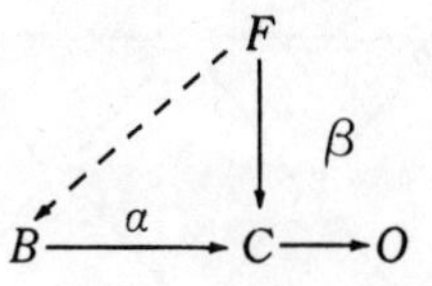

.

i.e., α is epic. If F is free, there is a map $\gamma: F \to B$ making the diagram commute.

Proof: Let $X = \{x_k: k \in K\}$ be a basis of F. Since α is epic, for each k there is an element $b_k \in B$ with $\alpha(b_k) = \beta(x_k)$. By the axiom of choice, there is a function $X \to B$ with $x_k \mapsto b_k$. Since F is free on X, there is a homomorphism $\gamma: F \to B$ with $\gamma x_k = b_k$. To see that $\alpha\gamma = \beta$, it suffices to check each on X, and $\alpha\gamma(x_k) = \alpha b_k = \beta x_k$.

COROLLARY 3.7. *If F is free, then the functor* Hom$(F, \quad)$ *is exact.*

Proof: Since Hom$(F, \quad)$ is always left exact, we need only show that if $0 \to A \xrightarrow{\alpha} B \xrightarrow{\beta} C \to 0$ is exact, then

$$\text{Hom}(F, B) \longrightarrow \text{Hom}(F, C) \longrightarrow 0$$

is exact, i.e., if $f \in \text{Hom}(F, C)$, then $f = \beta g$ for some $g \in \text{Hom}(F, B)$. But this is exactly the situation of Theorem 3.6:

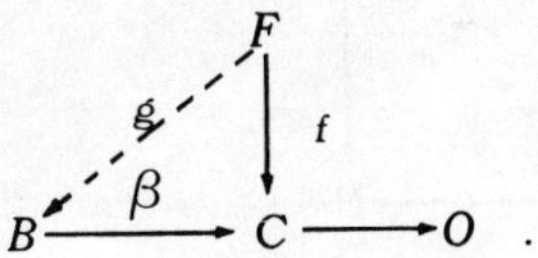

DEFINITION. A module P is **projective** if it behaves as F does in Theorem 3.6, i.e.,

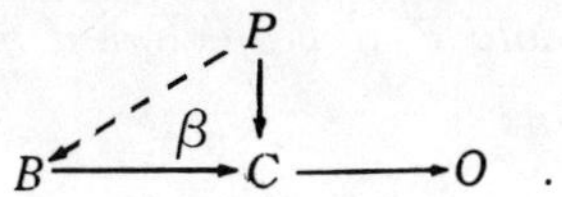

THEOREM 3.8. *A module P is projective if and only if* Hom$(P, \quad)$ *is exact.*

Proof: We have already seen (in Corollary 3.7) that if P is projective, then $\mathrm{Hom}(P,\ \ )$ is exact.

Suppose, conversely, that $\mathrm{Hom}(P,\ \ )$ is exact, and we have the situation

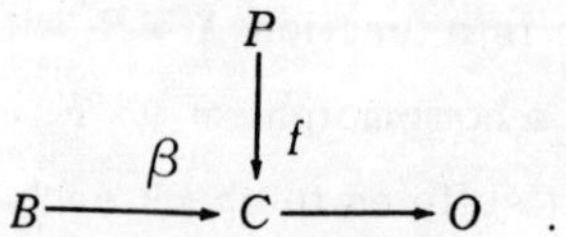

Now $\mathrm{Hom}(P,B) \to \mathrm{Hom}(P,C)$ is epic, so that there is $g \in \mathrm{Hom}(P,B)$ with $\beta g = f$, and so P is projective.

Is every projective module free? Let us first examine projectives a bit more.

THEOREM 3.9. *Let P be projective and let $\beta\colon B \to P$ be epic. Then $B = \ker \beta \oplus P'$, where $P' \cong P$.*

Proof: Consider the diagram:

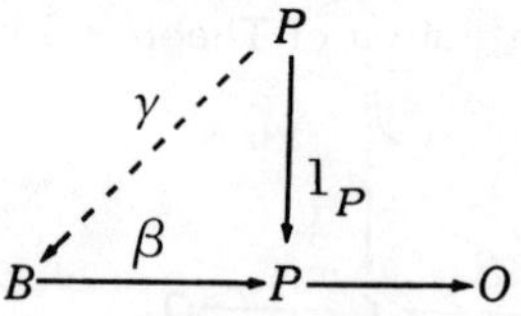

Since P is projective, there is a map $\gamma\colon P \to B$ with $\beta\gamma = 1_P$. By set theory, γ is monic, and so Theorem 2.5 gives the theorem.

THEOREM 3.10. *A module P is projective if and only if it is a summand of a free module.*

Proof: Every module, P included, is a quotient of a free module. There is thus an epic $\beta\colon F \to P$ for some free F. By Theorem 3.9, P projective gives P a summand of F.

Suppose, conversely, that P is a summand of a free module F. There are maps

$$P \underset{p}{\overset{i}{\rightleftarrows}} F$$

with $pi = 1_P$. Suppose now we have the situation

$$\begin{array}{ccccc} F & \overset{i}{\longleftarrow} & P & & \\ & \underset{p}{\longrightarrow} & & & \\ {\scriptstyle\gamma}\big\downarrow & & \big\downarrow{\scriptstyle f} & & \\ B & \overset{\beta}{\longrightarrow} & C & \longrightarrow & O \end{array}$$

Since F is projective, there is a map $\gamma: F \to B$ with $\beta\gamma = fp$. Define $\delta: P \to B$ by $\delta = \gamma i$. The diagram commutes, for

$$\beta\delta = \beta\gamma i = fpi = f \ .$$

Note that this argument shows that any summand of a projective is projective.

We can now give an example of a projective module that is not free. Let $R = Z_6$. Now $R = Z_2 \oplus Z_3$, so that Z_2 is R-projective, being a summand of the free module R. But Z_2 is not free, for any non zero free module has at least 6 elements. There are also examples of domains R having nonfree projectives.

THEOREM 3.11. *A module P is projective if and only if every exact sequence*

$$O \longrightarrow A \longrightarrow B \longrightarrow P \longrightarrow O$$

splits.

Proof: If P is projective, this is just Theorem 3.9. Conversely, a particular such sequence exists in which B is free so that P is a summand of a free and hence projective.

THEOREM 3.12. *If* $\{P_k : k \in K\}$ *is a family of projective modules, then* $\Sigma_{k \in K} P_k$ *is also projective.*

Proof: Consider the situation

$$\begin{array}{ccccc} & & \Sigma P_k & & \\ & & \downarrow{\scriptstyle f} & & \\ B & \xrightarrow{\beta} & C & \longrightarrow & O \end{array} .$$

If $i_k : P_k \to \Sigma P_k$ is injection, then $fi_k : P_k \to C$, so that projectivity of P_k gives a map $\gamma_k : P_k \to B$ with $\beta\gamma_k = fi_k$. By Theorem 2.1, the maps γ_k determine a map $\gamma: \Sigma P_k \to B$, and the reader may check $\beta\gamma = f$.

A product of projectives need not be projective; Baer proved that if $R = Z$, then ΠZ is not free when the index set is infinite.

Exercises 3.2. A *complement* of a projective module P is a module Q (necessarily projective) with $P \oplus Q$ free. Prove that every f.g. projective module has a f.g. complement.

3.3 (Eilenberg). Every projective module has a free complement. (Hint: Try to prove this for $R = Z_6$ and $P = Z_2$.)

We now dualize the definition of projective.

DEFINITION. A module E is **injective** in case, for any module B and any submodule A of B, every map $f: A \to E$ can be extended to a map $g: B \to E$. The diagram is

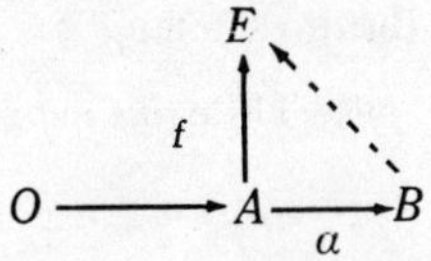

There are no immediate examples of injective modules, but we shall see that they exist in abundance. We now give duals for injective modules of the theorems we have just proved for projectives.

THEOREM 3.13. *A module* E *is injective if and only if the functor* Hom($, E)$ *is exact.*

Proof: If E is injective, it suffices to prove that if

$$0 \longrightarrow A \xrightarrow{\alpha} B \xrightarrow{\beta} C \longrightarrow 0$$

is exact, then

$$\mathrm{Hom}(B, E) \to \mathrm{Hom}(A, E)$$

is epic, i.e., if $f \in \mathrm{Hom}(A, E)$, there is $g \in \mathrm{Hom}(B, E)$ with $g\alpha = f$. But this is just the definition of injective.

The converse is just as easy and is left to the reader.

THEOREM 3.14. *Let* $\{E_k : k \in K\}$ *be a family of modules. Then* $\Pi\, E_k$ *is injective if and only if each* E_k *is injective.* (Recall that the dual of sum is product.)

Proof: Consider the diagram

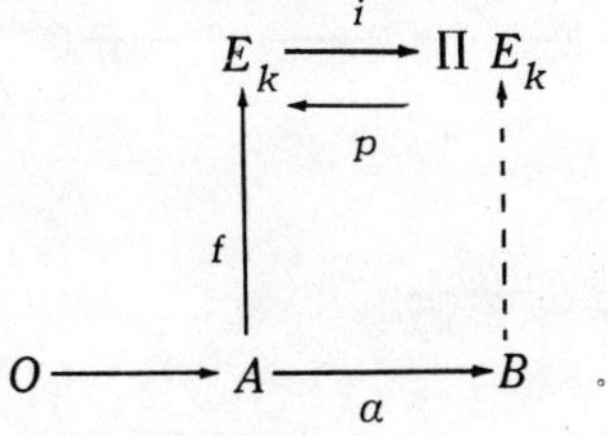

Since ΠE_k is injective, there is a map $g\colon B \to \Pi E_k$ with $ga = if$. Define $h\colon B \to E_k$ by $h = pg$. Then $ha = pga = pif = f$, as desired. Hence E_k is injective.

Suppose, conversely, that each E_k is injective, and let

$$E_k \underset{p_k}{\overset{i_k}{\rightleftarrows}} \Pi E_k$$

be the projections and injections. Consider the diagram

$$\begin{array}{ccccc} & & \Pi E_k & \underset{i_k}{\overset{p_k}{\rightleftarrows}} & E_k \\ & & {\scriptstyle f}\big\uparrow & & \big\uparrow \\ O & \longrightarrow & A & \xrightarrow[a]{} & B \end{array} .$$

Since E_k is injective, there is a map $g_k\colon B \to E_k$ with $g_k a = p_k f$. Define $h\colon B \to \Pi E_k$ by $h(b) = (g_k b)$. Then

$$haa = (g_k aa) = (p_k fa) = fa ,$$

so that $ha = f$ and ΠE_k is injective.

We shall see later that a sum of injectives need not be injective.

THEOREM 3.15. *If a module E is injective, then every exact sequence*

$$O \longrightarrow E \xrightarrow{i} B \longrightarrow C \longrightarrow O$$

splits.

Proof: Consider the diagram

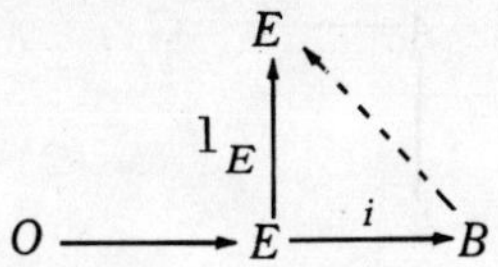

Since E is injective, there is a map $p\colon B \to E$ with $pi = 1_E$.

There would be no difficulty in proving the converse of Theorem 3.15 if we had the dual of Theorem 3.3: every module is a submodule of an injective. This is, in fact, true, but we will have to work a little to get it. It may be of interest to give a proof avoiding the imbedding into injectives. Besides, the simple construction we use will be needed elsewhere.

Suppose we have a diagram

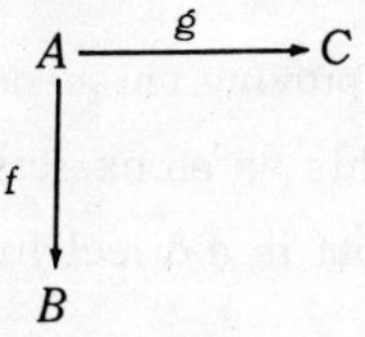

We wish to construct a module and maps so that we get a commutative diagram

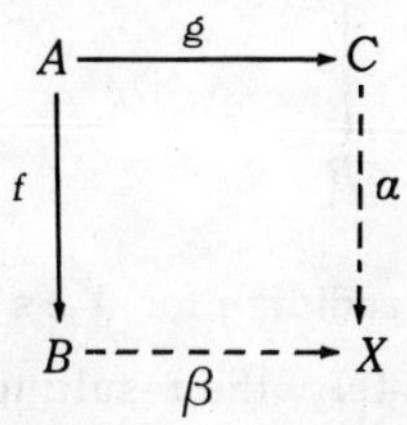

Call (X, α, β) a *solution* to this problem. Solutions are not unique: if (X, α, β) is a solution, and if $h\colon X \to Y$ is any map, then $(Y, h\alpha, h\beta)$ is also a solution.

DEFINITION. Given

$$\begin{array}{ccc} A & \xrightarrow{g} & C \\ {\scriptstyle f}\downarrow & & \\ B & & \end{array}$$

a **pushout** is a commutative diagram

$$\begin{array}{ccc} A & \xrightarrow{g} & C \\ {\scriptstyle f}\downarrow & & \downarrow{\scriptstyle \alpha} \\ B & \xrightarrow[\beta]{} & X \end{array}$$

which is largest in the sense that if (X', α', β') is another solution, then there is a unique map $h\colon X \to X'$ with $h\alpha = \alpha'$ and $h\beta = \beta'$.

There is no difficulty proving uniqueness of pushouts (to isomorphism), and we leave this as an exercise to the reader. A second exercise: show that pushout is a direct limit.

THEOREM 3.16. *A pushout diagram exists for*

$$\begin{array}{ccc} A & \xrightarrow{g} & C \\ {\scriptstyle f}\downarrow & & \\ B & & \end{array}$$

Proof: An obvious candidate for X is $B \oplus C$, with α and β the injections. Unfortunately, the resulting diagram need not commute. So, we make it commute. Let S be the submodule of $B \oplus C$ consisting of all elements of the form $(fa, -ga)$. Define $X = (B \oplus C)/S$, $\alpha c = (o, c) + S$ and $\beta b = (b, o) + S$. It is easily checked that (X, α, β) is a solution.

Suppose (X', α', β') is another solution. Define $h: X \to X'$ by $(b, c) + S \mapsto \alpha' b + \beta' c$. We let the reader check that h is well-defined and that we have constructed a pushout.

One may dualize this construction. Starting with

$$\begin{array}{ccc} & & B \\ & & \downarrow f \\ C & \xrightarrow{g} & A \end{array},$$

a **pullback** is a commutative square which is minimal in the sense that any other solution maps uniquely into it.

Exercises 3.4. Prove the existence of pullback diagrams (Hint: just dualize what we did above, so that you will select for X a submodule of $B \oplus C$ instead of a quotient. Alternatively, prove pullback is an inverse limit.)

3.5 If both f and g are monic, we may regard C and B as being submodules of A. Show that the pullback is just $C \cap B$.

3.6. Every left exact functor preserves pullbacks. In particular, if F is left exact and $A_1, A_2 \subset B$, then $F(A_1 \cap A_2) = FA_1 \cap FA_2$.

LEMMA 3.17. *Let the following diagram be a pushout.*

$$\begin{array}{ccc} A & \xrightarrow{g} & C \\ {\scriptstyle f}\downarrow & & \downarrow{\scriptstyle \alpha} \\ B & \xrightarrow{\beta} & X \end{array}$$

If g is monic, then β is monic.

Proof: Suppose $\beta b = O$. Now $\beta b = (b, o) + S$, so that $(b, o) \in S$;

therefore, $(b, o) = (fa, -ga)$ for some $a \in A$. Hence $b = fa$ and $O = ga$. Since g is monic, $a = O$, and so $b = fa = O$. It follows that β is monic.

THEOREM 3.18. *If every exact sequence*

$$O \longrightarrow E \xrightarrow{i} B \longrightarrow C \longrightarrow O$$

splits, then E is injective.

Proof: Consider the situation

$$\begin{array}{ccccc} & & E & & \\ & & \uparrow f & & \\ O & \longrightarrow & A & \xrightarrow{g} & M \end{array}$$

Construct the pushout diagram (both arrows emanate from A)

$$\begin{array}{ccccc} & & E & \xrightarrow{\beta} & X \\ & & f\uparrow & & \uparrow\alpha \\ O & \longrightarrow & A & \xrightarrow{g} & M \end{array}$$

By the lemma, β is monic. By hypothesis, E is a summand of X, i.e., there is a map $p: X \to E$ with $p\beta = 1_E$. Define $\gamma: M \to E$ by $\gamma = p\alpha$. Then $\gamma g = p\alpha g = p\beta f = f$, and so E is injective.

Again, this proof of Theorem 3.18 is just a divertissement. A quick proof will be available as soon as we show there are plenty of injectives. We remind the reader that we don't yet have any examples of injectives!

THEOREM 3.19. *A module E is injective if and only if every map $f: I \to E$, where I is a left ideal of R, can be extended to R.*

Proof: Suppose E is injective. A left ideal is just a submodule of R, so this is just a special case of the definition of injective.

Now suppose we have the situation

$$\begin{array}{ccccc} & & E & & \\ & & \uparrow{\scriptstyle f} & & \\ 0 & \longrightarrow & A & \longrightarrow & B \end{array} \quad .$$

For notational convenience, we assume A is a submodule of B. Let $\mathfrak{F}$ consist of all pairs (A', f'), where $A \subset A' \subset B$ and f': $A' \to E$ extends f; $\mathfrak{F}$ is nonempty, for $(A, f) \in \mathfrak{F}$. Partially order $\mathfrak{F}$ by saying $(A', f') \leq (A'', f'')$ in case $A' \subset A''$ and f'' extends f'. By Zorn's Lemma, there is a maximal pair (A_o, f_o). If $A_o = B$, we are done.

Suppose $A_o \neq B$, and let $x \in B - A_o$. If $I = \{r \in R\colon rx \in A_o\}$, then I is a left ideal of R. Define $h\colon I \to E$ by $r \mapsto f_o(rx)$. By hypothesis, there is a map $t\colon R \to E$ extending h. Define $A_1 = A_o + Rx$, and define $f_1\colon A_1 \to E$ by $a_o + rx \mapsto f_o(a_o) + rt(1)$. First, f_1 is well-defined. Suppose $a_o + rx = a'_o + r'x$. Then $(r - r')x = a'_o - a_o \in A_o$, so we may apply f_o to get

$$f_o(a'_o - a_o) = f_o((r-r')x) = h(r-r') = t(r-r') = (r-r')t(1).$$

Therefore $f_o a' + r't(1) = f_o a_o + rt(1)$, as desired. Second, f_1 does extend f_0: $f_1 a_o = f_o a_o$. The pair (A_1, f_1) thus lies in $\mathfrak{F}$ and it is larger than the maximal pair (A_o, f_o). This contradiction shows that $A_o = B$, so $f_o\colon B \to E$ and E is injective.

DEFINITION. Let R be a ring, A an R-module. If $r \in R$ and $a \in A$, we say that *a is divisible by r* if $ra' = a$ for some $a' \in A$;

we say that A is **divisible** if each $a \in A$ is divisible by every non-zero-divisor r in R.

EXAMPLE. The additive group of rationals, Q, is a divisible abelian group (= Z-module).

THEOREM 3.20. *Every injective module E is divisible.*

Proof: Let $a \in E$ and let $r_o \in R$ be a non-zero-divisor. Define $f\colon Rr_o \to E$ by $rr_o \mapsto ra$; note that f is well-defined because r_o is not a zero divisor. Since E is injective, there is a map $h\colon R \to E$ which extends f. In particular,

$$a = f(r_o) = h(r_o) = r_o h(1) ,$$

so that a is divisible by r_o.

THEOREM 3.21. *Every quotient A of a divisible module D is also divisible.*

Proof: Suppose $f\colon D \to A$ is epic, $a \in A$, and $r \in R$ is not a zero divisor. Lift a to $d \in D$, i.e., $d \mapsto a$. Since D is divisible, $d = rd'$ for some $d' \in D$, so that $a = f(d) = f(rd') = rf(d')$; therefor A is divisible.

We remark that the converse of Theorem 3.20 is false: there are divisible modules that are not injective. We shall see that a quotient of an injective need not be injective (for some rings R), so that Theorem 3.21 would then provide an example.

Exercises 3.7. Let R be a domain and A a torsion-free R-module (A is **torsion free** if $r \in R$, $a \in A$, and $ra = O$ implies $r = O$ or $a = O$). Prove that A is injective if and only if A is divisible.

3.8. Let R be a domain with quotient field Q. A torsion-free R-module is divisible if and only if it is a vector space over Q.

3.9. Let $\{A_k : k \in K\}$ be modules. Then ΣA_k is divisible if and only if each A_k is divisible. Similarly, if each A_k is divisible, so is ΠA_k.

For some rings R, divisible implies injective.

THEOREM 3.22. *Let R have no zero divisors and suppose every left ideal is principal. Then every divisible R-module E is injective.*

Proof: Let I be a nonzero left ideal of R. By Theorem 3.19, it suffices to complete the diagram

$$\begin{array}{ccccc} & & E & & \\ & & \uparrow{\scriptstyle f} & & \\ O & \longrightarrow & I & \longrightarrow & R \end{array} .$$

By hypothesis, $I = Rr_o$ for some $r_o \in R$; moreover, r_o is not a zero divisor. Since E is divisible, there is an element $a \in E$ with $r_o a = f(r_o)$. Define $h: R \to E$ by $r \mapsto ra$, and observe that h extends f.

COROLLARY 3.23. *If R is a principal ideal domain, an R-module is injective if and only if it is divisible.*

We finally have some examples of injectives, over a restricted class of rings, we admit. Later we shall see that Corollary 3.23 can be improved by replacing principal ideal domains by a larger class of rings, Dedekind rings.

Here is a special case of the imbedding theorem we seek.

THEOREM 3.24. *Every abelian group G can be imbedded in an injective group.*

Proof: Write $G = F/S$, where F is free. Thus $F = \Sigma\, Z$. If we imbed each copy of Z in a copy of the rationals, Q, we have

$$G = F/S = (\Sigma\, Z)/S \subset (\Sigma\, Q)/S\ .$$

Since Q is divisible, $\Sigma\, Q$ is divisible, and $(\Sigma\, Q)/S$ is divisible. By Corollary 3.23, $(\Sigma\, Q)/S$ is injective.

THEOREM 3.25. *If D is a divisible abelian group, then* $\mathrm{Hom}_Z(R, D)$ *is an injective left R-module.*

Proof: We make $\mathrm{Hom}_Z(R, D)$ into a left R-module in precisely the same way described in our preamble to the adjoint isomorphism 2.8: if $f\colon R \to D$, define rf by $r' \mapsto f(r'r)$.

We must show $\mathrm{Hom}_R(\ \ , \mathrm{Hom}_Z(R, D))$ is exact, i.e., it converts monomorphisms to epimorphisms. The adjoint isomorphism gives a commutative diagram

$$\begin{array}{ccc}
\mathrm{Hom}_R(B, \mathrm{Hom}_Z(R, D)) & \longrightarrow & \mathrm{Hom}_R(A, \mathrm{Hom}_Z(R, D)) \\
\downarrow & & \downarrow \\
\mathrm{Hom}_Z(B, D) & \longrightarrow & \mathrm{Hom}_Z(A, D)
\end{array}$$

whenever $0 \to A \to B$ is an exact sequence of R-modules, where we have identified $R \otimes_R B$ with B and $R \otimes_R A$ with A. Since D is a divisible group, it is Z-injective, and so the bottom map is epic, by Theorem 3.13. It follows that the top map is epic too. Therefore $\mathrm{Hom}_Z(R, D)$ is an injective module.

REMARK. We have proved a little more: if R is an S-algebra and D is an injective S-module, then $\mathrm{Hom}_S(R, D)$ is an injective R-module.

THEOREM 3.26. *Every R-module A can be imbedded in an injective R-module.*

Proof: Considering A only as an abelian group, there is a Z-monomorphism

$$O \longrightarrow A \xrightarrow{i} D$$

where D is a suitable divisible group. We have

$$A \cong \mathrm{Hom}_R(R, A) \to \mathrm{Hom}_Z(R, A) \to \mathrm{Hom}_Z(R, D) ;$$

these maps are: $a \mapsto f$, where $f(1) = a$; regard an R-map $f: R \to A$ as a Z-map; if $g \in \mathrm{Hom}_Z(R, A)$, then $g \mapsto ig$. The first map is an isomorphism, the second map is visibly monic, and the third map is monic because Hom is left exact.

Conclusion: the composite $A \to \mathrm{Hom}_Z(R, D)$ is monic. One must also check that this map is, in fact, an R-map; it is because each of the three pieces is (brief exercise).

Exercise 3.10. Use Theorem 3.26 to prove that if A has the property that every short exact sequence $O \to A \to B \to C \to O$ splits, then A is injective.

DEFINITION. Given a module A, an **injective envelope** of A is a module E containing it such that

i) E is injective;

ii) there is no injective module M properly between A and E.

One may prove (Curtis-Reiner, p. 390) that every module A has an injective envelope $E(A)$ that is unique to isomorphism; it is an **essential extension** of A in the sense that every nonzero submodule of $E(A)$ meets A in a nonzero submodule. (The correspondence $A \to E(A)$ cannot be extended to define a functor.)

DEFINITION. An **injective resolution** of a module A is an exact sequence

$$0 \longrightarrow A \longrightarrow E_o \longrightarrow E_1 \longrightarrow \cdots$$

where each E_n is injective.

THEOREM 3.27. *Every module has an injective resolution.*

Proof: Dual to Theorem 3.5.

Observe that an injective resolution is a complex (if we consider 0's going off to the left). Earlier, we defined a free resolution; one may also define projective resolutions in the obvious way.

In the light of Theorems 3.8 and 3.13, the following definition is quite reasonable.

DEFINITION. A right R-module B is **flat** if the functor $B \otimes_R$ is an exact functor.

Note that B is flat if and only if, whenever f is a monomorphism, then $1_B \otimes f$ is a monomorphism.

EXAMPLES 1: R itself is flat (see exercises 2.17 and 2.18).

2: If $R = Z$, we saw in Chapter 2 that torsion groups are not flat.

THEOREM 3.28. *Let* $\{B_k\colon k \in K\}$ *be a family of right* R*-modules. Then* $\Sigma\, B_k$ *is flat if and only if each* B_k *is flat.*

Proof: First, a remark on sums. Given a family of maps $f_k\colon A'_k \to A_k$, there is a unique map $\Sigma\, f_k\colon \Sigma\, A'_k \to \Sigma\, A_k$ with $\Sigma\, a'_k \mapsto \Sigma\, f_k a'_k$; furthermore, $\Sigma\, f_k$ is monic if and only if each f_k is monic. The proof of this is easy and is left to the reader.

Suppose $O \to A' \xrightarrow{f} A$ is exact. We have a commutative diagram

$$\begin{array}{ccc} (\Sigma B_k) \otimes A' & \xrightarrow{1 \otimes f} & (\Sigma B_k) \otimes A \\ \downarrow & & \downarrow \\ \Sigma (B_k \otimes A') & \xrightarrow{\Sigma(1_k \otimes f)} & \Sigma (B_k \otimes A) \end{array} ,$$

where the downward maps are the usual isomorphisms

$$(\Sigma\, b_k) \otimes a \mapsto \Sigma\, (b_k \otimes a) \quad \text{and} \quad 1_k \otimes f = 1_{B_k} \otimes f .$$

The preamble now gives $1 \otimes f$ monic if and only if each $1_k \otimes f$ is monic, i.e., $\Sigma\, B_k$ is flat if and only if each B_k is flat.

COROLLARY 3.29. Every projective module is flat.

Proof: Since R is flat, so is every free module, hence so is every summand of a free.

Direct limits are a useful tool in constructing flat modules.

THEOREM 3.30. *Let* $\{B_k, \phi_{kj}\}$ *be a direct system of modules in which the index set* K *is a directed set. If each* B_k *is flat, then* $\varinjlim B_k$ *is flat.*

Proof: If $O \to A' \xrightarrow{f} A$ is exact, then there is a commutative diagram

$$\begin{array}{ccc} (\varinjlim B_k) \otimes A' & \xrightarrow{1 \otimes f} & (\varinjlim B_k) \otimes A \\ \downarrow & & \downarrow \\ \varinjlim (B_k \otimes A') & \xrightarrow{\varinjlim (1_k \otimes f)} & \varinjlim (B_k \otimes A) \end{array} ,$$

where the downward maps are the isomorphisms of exercise 2.19: $(b_k + S) \otimes a \mapsto (b_k \otimes a) + S_o$ (where S, S_o are the relations divided out to give the direct limit), and the bottom map is

$$(b_k \otimes a') + S' \mapsto (b_k \otimes fa') + S$$

(the map obtained from a morphism between direct systems). By Theorem 2.13, $1 \otimes f$ is monic because each $1_k \otimes f = 1_{B_k} \otimes f$ is. Therefore $\varinjlim B_k$ is flat.

COROLLARY 3.31. *If every f.g. submodule of B is flat, then B is flat.*

Proof: We saw in exercise 2.13 that every module is a direct limit of its f.g. submodules.

COROLLARY 3.32. *If R is a domain, then its quotient field Q is a flat R-module.*

Proof: We saw in exercise 2.14 that Q is a direct limit of cyclic modules (each of which is isomorphic to R).

It follows that the converse of 3.29 need not be true: there may be flat modules that are not projective. For example, if $R = Z$, then Q = rationals, and it is easy to check that Q is not a summand of a free abelian group (it is not even isomorphic to a subgroup of a free abelian group).

DEFINITION: If B is a right R-module, its **character module,** B^*, is the *left* R-module $\mathrm{Hom}_Z(B, Q/Z)$, where the action of R is defined by $(rf)b = f(br)$.

(Q/Z is the rationals modulo one, i.e., rationals modulo integers, which we know is an injective Z-module, by Corollary 3.23.)

The next lemma shows why Q/Z was chosen in the definition of character module.

LEMMA 3.33. *Let A be an abelian group with subgroup A', and let $a \in A$, $a \notin A'$. Then there is a map $f\colon A \to Q/Z$ with $f(A') = 0$ and $fa \neq 0$.*

Proof: First note, for any positive integer n, that Q/Z contains an element of order n, namely, $1/n + Z$. It follows that, given any abelian group G and any nonzero $x \in G$, there is a map $h\colon Zx \to Q/Z$ with $hx \neq 0$ (Zx is the cyclic subgroup generated by x). Since Q/Z is injective, there is a map $g\colon G \to Q/Z$ that extends h; in particular, $gx \neq 0$. To complete the proof, take $G = A/A'$, $x = a + A'$, and f the composite of g with the natural map $A \to A/A'$.

The next lemma shows that character modules give a simple criterion for exactness (compare Lemma 2.10).

LEMMA 3.34. *A sequence of right R-modules*

$$0 \longrightarrow A \xrightarrow{\alpha} B \xrightarrow{\beta} C \longrightarrow 0$$

is exact if and only if the sequence of character modules

$$0 \longrightarrow C^* \xrightarrow{\beta^*} B^* \xrightarrow{\alpha^*} A^* \longrightarrow 0$$

is exact.

Proof: If the original sequence is exact, then the exact functor $\mathrm{Hom}_Z(\ \ , Q/Z)$ carries it into an exact sequence.

To prove the converse, we shall show that $\ker \alpha^* = \mathrm{im}\, \beta^*$ implies $\ker \beta = \mathrm{im}\, \alpha$, and we shall not assume α monic or β epic. The result will then follow.

Is $\mathrm{im}\, \alpha \subset \ker \beta$? Suppose $a \in A$ and $\alpha a \notin \ker \beta$, i.e., $\beta\alpha a \neq 0$. By Lemma 3.33, there is a map $g \in C^*$ with $g\beta\alpha a \neq 0$. But $g\beta\alpha = \alpha^*\beta^*(g) = 0$, since $\mathrm{im}\, \beta^* \subset \ker \alpha^*$.

Is $\ker \beta \subset \operatorname{im} \alpha$? If not, there is an element $x \in \ker \beta$, $x \notin \operatorname{im} \alpha$. By Lemma 3.33, there is a map $f \in B^*$ with $f(\operatorname{im} \alpha) = 0$ and $fx \neq 0$. Now $f(\operatorname{im} \alpha) = 0$ says $O = f\alpha = \alpha^*(f)$, so that $f \in \ker \alpha^* \subset \operatorname{im} \beta^*$. Hence $f = \beta^*(g) = g\beta$, for some $g \in C^*$. In particular, $fx = g\beta x = O$ since $x \in \ker \beta$, contradicting the fact that $fx \neq O$.

THEOREM 3.35. *A right R-module B is flat if and only if its character module B^* is injective as a left R-module.*

Proof: Suppose B is flat. Exactness of $O \to A' \to A$ (left modules) gives exactness of $O \to B \otimes A' \to B \otimes A$, which, in turn, gives exactness of $(B \otimes A)^* \to (B \otimes A')^* \to O$. Observe that B is a 2-sided module: $B = {}_Z B_R$. The adjoint isomorphism 2.8 gives a commutative diagram

$$\begin{array}{ccccc} \operatorname{Hom}_Z(B \otimes A, Q/Z) & \longrightarrow & \operatorname{Hom}_Z(B \otimes A', Q/Z) & \longrightarrow & O \\ \downarrow & & \downarrow & & \\ \operatorname{Hom}_R(A, B^*) & \longrightarrow & \operatorname{Hom}_R(A', B^*) & \longrightarrow & O \end{array} .$$

It follows that the bottom row is exact, i.e., $\operatorname{Hom}_R(\ \ , B^*)$ is an exact functor. Therefore, B^* is an injective R-module.

Suppose now that B^* is an injective left R-module, and $O \to A' \to A$ is an exact sequence of left R-modules. We have a commutative diagram

$$\begin{array}{ccccc} \operatorname{Hom}_R(A, B^*) & \longrightarrow & \operatorname{Hom}_R(A', B^*) & \longrightarrow & O \\ \downarrow & & \downarrow & & \\ \operatorname{Hom}_Z(B \otimes_R A, Q/Z) & \longrightarrow & \operatorname{Hom}_Z(B \otimes_R A', Q/Z) & \longrightarrow & O \end{array} ,$$

so that exactness of the top row gives exactness of the bottom, i.e., $(B \otimes A)^* \to (B \otimes A')^* \to O$ is exact. By Lemma 3.34 $O \to B \otimes A' \to B \otimes A$ is exact, and so B is flat.

THEOREM 3.36. *Let B be a right R-module. If, for each f.g. left ideal I of R, $O \to B \otimes I \xrightarrow{1 \otimes i} B \otimes R$ is exact (where $i: I \to R$ is inclusion), then B is flat.*

(Note that the hypothesis merely says that if I is any f.g. left ideal of R, $B \otimes I$ can be identified with BI via $b \otimes i \mapsto bi$.)

Proof: Since every left ideal is a direct limit of f.g. left ideals, Theorem 2.13 implies $O \to B \otimes I \to B \otimes R$ is exact for every left ideal I of R. As in the proof of Theorem 3.35, we have exactness of

$$\mathrm{Hom}_R(R, B^*) \longrightarrow \mathrm{Hom}_R(I, B^*) \longrightarrow O$$

for every left ideal I. By Theorem 3.19, this implies B^* is injective, so that B is flat, by Theorem 3.35.

The final assertion follows from observing that if $\tau: B \otimes R \to B$ is the isomorphism $b \otimes r \mapsto br$, then $\tau(1 \otimes i)$ is monic.

We know that a quotient of a flat module need not be flat (for free modules are flat). The answer to the question which quotients are flat shows that flatness is related to the notion of "purity" for abelian groups.

THEOREM 3.37. *Suppose $O \to K \to F \xrightarrow{\beta} B \to O$ is an exact sequence of right R-modules, where F is flat. If B is flat, then $K \cap FI = KI$ for every left ideal I of R. Conversely, if $K \cap FI = KI$ for every f.g. left ideal I of R, then B is flat.*

Proof: Observe that $KI \subset K \cap FI$ always, so it is only the reverse inclusion that is significant.

Tensoring the original sequence by a left ideal I gives exactness of

$$K \otimes I \to F \otimes I \xrightarrow{\beta \otimes 1} B \otimes I \to 0 \ .$$

By Theorem 3.36, we may identify $F \otimes I$ with FI via $f \otimes i \mapsto fi$; moreover, image $(K \otimes I)$ is then identified with KI. There results an isomorphism

$$FI/KI \to B \otimes I$$

given by $fi + KI \mapsto \beta f \otimes i$.

Observe that $BI = \{\Sigma_j (\beta f_j) i_j : f_j \in F, \ i_j \in I\}$

$$= \{\beta(\Sigma_j f_j i_j)\} = \beta(FI) \ .$$

There is thus the Noether isomorphism

$$BI = \beta(FI) \longrightarrow FI/FI \cap K \, ,$$

given by $bi \mapsto fi + FI \cap K$, where $\beta f = b$.

We have the composite σ

$$FI/KI \longrightarrow B \otimes I \xrightarrow{\theta} BI \to FI/FI \cap K \, ,$$

where $\theta(b \otimes i) = bi$. Explicitly, σ takes $fi + KI \mapsto fi + FI \cap K$. Since $KI \subset K \cap FI$, σ is an isomorphism if and only if $KI = FI \cap K$. Moreover, since the flanking maps are isomorphisms, σ is an isomorphism if and only if θ is.

If B is flat, θ is an isomorphism for all I and $KI = FI \cap K$. Conversely, if $KI = FI \cap K$ for all f.g. left ideals I, then θ is an isomorphism for all f.g. ideals I, and B is flat.

Exercise 3.11. If R is a domain, every flat module is torsion free.

LEMMA 3.38. *Let $0 \to K \to F \to B \to 0$ be an exact sequence of right R-modules, where F is free with basis $\{x_j : j \in J\}$.* If $v = x_{j_1}r_1 + \cdots + x_{j_t}r_t$ is an element of F, define I_v to be the left ideal of R generated by $r_1, \ldots, r_t$. Then B is flat if and only if, for every $v \in K$, we have $v \in KI_v$.

Proof: If B is flat and $v \in K$, then $v \in K \cap FI_v = KI_v$, by Theorem 3.37. Conversely, let I be any left ideal of R, and let $v \in K \cap FI$. Then $I_v \subset I$ and $v \in K \cap FI_v$, so that $v \in KI_v \subset KI$. Hence $K \cap FI = KI$ and B is flat, by Theorem 3.37.

THEOREM 3.39 (Villamayor). *Let $0 \to K \to F \to B \to 0$ be an exact sequence of right R-modules, where F is free. The following are equivalent;*

(i) *B is flat;*

(ii) *For every $v \in K$, there is a map $\theta : F \to K$ with $\theta v = v$;*

(iii) *For every $v_1, \ldots, v_n \in K$, there is a map $\theta : F \to K$ with $\theta v_i = v_i$, $i = 1, \ldots, n$.*

Proof: (i) $\Longrightarrow$ (ii) Suppose B is flat and $v \in K$. Then $v = x_{j_1}r_1 + \cdots + x_{j_t}r_t$ and $I_v = Rr_1 + \cdots + Rr_t$. Since $v \in K \cap FI_v$, Lemma 3.38 gives $v \in KI_v$, i.e., $v = \Sigma\, k_\lambda \sigma_\lambda$, $k_\lambda \in K$, $\sigma_\lambda \in I_v$. Now each $\sigma_\lambda = \Sigma_i\, s_{\lambda i} r_i$, so that $v = \Sigma\, k_i' r_i$ (where $k_i' = \Sigma\, k_\lambda s_{\lambda i} \in K$). Define $\theta : F \to K$ by $\theta(x_{j_i}) = k_i'$, while θ sends all the other basis elements of F into 0.

(ii) $\Longrightarrow$ (i) Suppose $v = f_1\sigma_1 + \cdots + f_m\sigma_m \in K \cap FI_v$, and $\theta : F \to K$ with $\theta v = v$. Then $v = \theta v = \theta(f_1)\sigma_1 + \cdots + \theta(f_m)\sigma_m \in KI_v$, so that B is flat, by Lemma 3.38.

Since (iii) trivially implies (ii), it only remains to prove (ii) $\Longrightarrow$ (iii). We do an induction on n, the case $n = 1$ being our assumption (ii). Suppose $n > 1$ and we have elements $v_1, \ldots, v_h \in K$. By (ii), there is a map $\theta_n : F \to K$ with $\theta_n v_n = v_n$. Define $v_i' \in K$ by $v_i' = v_i - \theta_n v_i$, $i = 1, \ldots, n-1$, By induction, there is a map $\theta' : F \to K$ with $\theta' v_i' = v_i'$, $i = 1, \ldots, n-1$. Finally, define $\theta = 1_F - (1_F - \theta') \circ (1_F - \theta_n)$. It is easy to check that $\operatorname{im} \theta \subset K$ and $\theta v_i = v_i$, all i.

DEFINITION. A module B is **finitely related** if there is an exact sequence

$$O \longrightarrow K \longrightarrow F \longrightarrow B \longrightarrow O$$

where F is free and both F and K are finitely generated.

COROLLARY 3.40. *A finitely related flat module B is projective.*

Proof: There is an exact sequence

$$O \longrightarrow K \xrightarrow{\ i\ } F \longrightarrow B \longrightarrow O$$

where F is free and K is generated by a finite number of elements, say, $v_1, \ldots, v_n$. By Theorem 3.39, there is a map $\theta : F \to K$ with $\theta i = 1_k$, so that the sequence splits and B is projective.

Here is an example of a f.g. module that is not finitely related. Let R be a ring having an ideal I that cannot be finitely generated, e.g., $R = F[x_1, x_2, \ldots] =$ polynomials in infinitely many variables, and $I =$ the ideal generated by the indeterminates. Then R/I is cyclic, but not finitely related, as we shall see from the next theorem.

THEOREM 3.41 (Schanuel's Lemma). *Consider the exact sequences*

$$0 \longrightarrow K_1 \longrightarrow P_1 \longrightarrow B \longrightarrow 0$$

and

$$0 \longrightarrow K_2 \longrightarrow P_2 \longrightarrow B \longrightarrow 0 \quad ,$$

where P_1 *and* P_2 *are projective. Then* $K_1 \oplus P_2 \cong K_2 \oplus P_1$.

Proof: Consider the diagram with exact rows

$$\begin{array}{ccccccccc} 0 & \longrightarrow & K_1 & \xrightarrow{i} & P_1 & \xrightarrow{p} & B & \longrightarrow & 0 \\ & & \downarrow{\scriptstyle \alpha} & & \downarrow{\scriptstyle \beta} & & \downarrow{\scriptstyle 1_B} & & \\ 0 & \longrightarrow & K_2 & \xrightarrow[j]{} & P_2 & \xrightarrow[q]{} & B & \longrightarrow & 0 \end{array}$$

Since P_1 is projective, there is a map $P_1 \xrightarrow{\beta} P_2$ making the square on the right commute; diagram chasing now gives a map $K_1 \xrightarrow{\alpha} K_2$ making the other square commute.

Given such a diagram, there is an exact sequence

$$0 \longrightarrow K_1 \xrightarrow{\theta} P_1 \oplus K_2 \xrightarrow{\psi} P_2 \longrightarrow 0 \ ,$$

where $k_1 \mapsto (ik_1, \alpha k_1)$ and $(p_1, k_2) \mapsto \beta p_1 - jk_2$. (The verification of exactness is routine.) If we now assume that P_2 is projective, this last sequence splits, i.e., $K_1 \oplus P_2 \cong K_2 \oplus P_1$, as desired.

COROLLARY 3.42. *If* B *is finitely related and*

$$0 \longrightarrow K \longrightarrow F \longrightarrow B \longrightarrow 0$$

is exact, where F *is a f.g. free module, then* K *is f.g.*

Proof: Since B is finitely related, there is some exact sequence

$$0 \longrightarrow K_1 \longrightarrow F_1 \longrightarrow B \longrightarrow 0$$

with F_1 free and both K_1 and F_1 f.g. By Schanuel's Lemma, $K_1 \oplus F \cong K \oplus F_1$. Thus K is an image (it's a summand) of the f.g. module $K_1 \oplus F$, and so K is f.g.

It now follows that the cyclic module exhibited above is not finitely related.

Given a left R-module B and a short exact sequence of right R-modules $0 \to A' \to A \to A'' \to 0$, we know that

$$0 \longrightarrow A' \otimes B \longrightarrow A \otimes B \longrightarrow A'' \otimes B \longrightarrow 0$$

need not be exact. Of course, if B is flat, the tensored sequence is exact. Instead of looking at B, we now focus on the sequence.

DEFINITION. An exact sequence of right R-modules,

$$0 \longrightarrow A' \xrightarrow{\lambda} A \longrightarrow A'' \longrightarrow 0,$$

is **pure exact** if, for every left R-module B, the sequence

$$0 \longrightarrow A' \otimes B \xrightarrow{\lambda \otimes 1} A \otimes B \longrightarrow A'' \otimes B \longrightarrow 0$$

is exact.

One also says that $\lambda A'$ is a **pure submodule** of A.

Exercises 3.12. Every summand of A is a pure submodule.

3.13. If a short exact sequence remains exact after tensoring by every finitely related module, then it is pure exact.

3.14. If $0 \to A' \to A \to A'' \to 0$ remains exact after applying $\operatorname{Hom}(B, \ \ )$, all B, then the sequence is split.

LEMMA 3.43. *Let B be a finitely related left R-module with generators $b_1, \ldots, b_n$ and relations $\sum r_{ij} b_j$, $i = 1, \ldots, m$. If, for a right R-module A, we have*

$$\sum_{j=1}^{n} a_j \otimes b_j = O \text{ in } A \otimes B,$$

then there are elements $\alpha_i \in A$ *with* $a_j = \sum \alpha_i r_{ij}$, *all* j.

Proof: There is an exact sequence

$$O \longrightarrow N \xrightarrow{\mu} F \xrightarrow{\phi} B \longrightarrow O \quad ,$$

where F is free on $v_1, \ldots, v_n$, N is the submodule of F generated by $\sum r_{ij} v_j$, $i = 1, \ldots, m$, and $\phi v_j = b_j$. Tensoring by A gives exactness of

$$A \otimes N \xrightarrow{1 \otimes \mu} A \otimes F \xrightarrow{1 \otimes \phi} A \otimes B \longrightarrow O \quad ,$$

and, by hypothesis, $\sum a_j \otimes v_j \in \ker 1 \otimes \phi = \operatorname{im} 1 \otimes \mu$. Now every element of $A \otimes N$ has an expression of the form

$$\sum_{i,j} \alpha_i \otimes r_{ij} v_j, \quad \text{where } \alpha_i \in A .$$

In particular,

$$\sum a_j \otimes v_j = (1 \otimes \mu) \sum_{i,j} \alpha_i \otimes r_{ij} v_j$$

$$= \sum_j \left(\sum_i \alpha_i r_{ij} \right) \otimes v_j .$$

Since F is free on the v_j's, every element of $A \otimes F$ has a unique expression of the form $\sum h_j \otimes v_j$, where $h_j \in A$. It follows that $a_j = \sum \alpha_i r_{ij}$, $j = 1, \ldots, n$.

THEOREM 3.44. (P. M. Cohn) *A submodule* $\lambda A'$ *of* A *is a pure submodule if and only if, whenever* $a'_j = \sum a_i r_{ij}$ *(where* $a'_j \in A'$,

$a_i \in A$, *and* $r_{ij} \in R$), *there are elements* $a_i' \in A'$ *with* $a_j' = \Sigma\, a_i' r_{ij}$.

REMARK. This condition can be translated into the following diagrammatic statement:

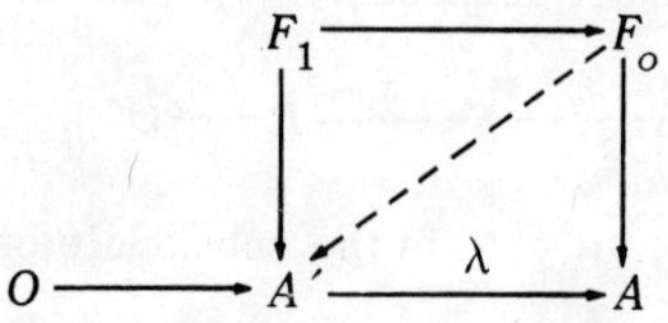

Given any commutative diagram with F_1, F_o f.g. free modules, there is a map $F_o \to A'$ making the new diagram commute.

Proof: Suppose $\lambda A'$ is a pure submodule of A, and $a_j' = \Sigma\, a_i r_{ij}$. Define a (finitely related) module $B = F/N$, where F is free on v_j, $j = 1,\ldots,n$, and N is generated by $\Sigma\, r_{ij} v_j$, $i = 1,\ldots,m$; let $b_j = v_j + N$. In $A \otimes B$, we have

$$\Sigma\, a_j' \otimes b_j = \sum_j \left(\Sigma\, a_i r_{ij} \otimes b_j \right)$$

$$= \sum_i \left(a_i \otimes \sum_j r_{ij} b_j \right) = O \;.$$

By purity, we have $\Sigma\, a_j' \otimes b_j = O$ in $A' \otimes B$; by the Lemma, there are elements $a_i' \in A'$ with $a_j' = \Sigma\, a_i' r_{ij}$.

For the converse, we must show that $O \to A' \otimes B \xrightarrow{\lambda \otimes 1} A \otimes B$ is exact for every module B. By exercise 3.13, we may assume B is finitely related, say, with generators $b_1, \ldots, b_n$ and relations $\Sigma\, r_{ij} b_j$, $i = 1,\ldots,m$. A typical element of $A' \otimes B$ can be written $\Sigma\, a_j' \otimes b_j$, where $a_j' \in A'$. If $(\lambda \otimes 1)\, \Sigma\, a_j' \otimes b_j = O$ in $A \otimes B$, is $\Sigma\, a_j' \otimes b_j = O$ in $A' \otimes B$? By the Lemma, there are elements

$a_i \in A$ with $a_j' = \sum a_i r_{ij}$; by hypothesis, there are elements $a_i' \in A'$ with $a_j' = \sum a_i' r_{ij}$. Therefore

$$\sum a_j' \otimes b_j = \sum_j \left(\sum a_i' r_{ij} \otimes b_j \right)$$

$$= \sum_i a_i' \otimes \sum r_{ij} b_j = O \text{ in } A' \otimes B \ .$$

Hence $\lambda \otimes 1$ is monic and $\lambda A'$ is a pure submodule of A.

Exercise 3.15. If $R = Z$, prove that A' is a pure subgroup of A if and only if $A' \cap nA = nA'$ for all $n \in Z$.

3.16. If A and B are torsion abelian groups, prove that $A \otimes B$ is a sum of cyclic groups. (Hint: Use Kulikov's Theorem [Fuchs, page 98]: if A is a torsion abelian group, there is a pure exact sequence $O \to \Sigma \to A \to D \to O$, where Σ is a sum of cyclic groups and D is divisible.)

REMARK: The diagram in Theorem 3.44 begs to be completed:

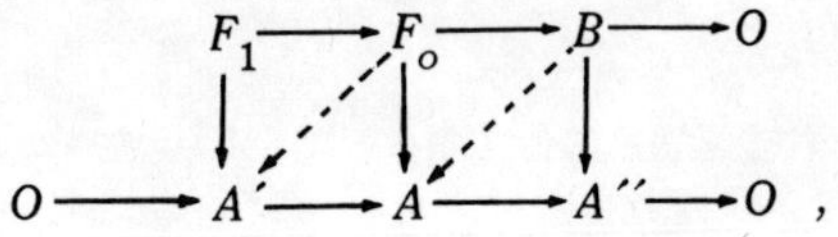

and the theorem it suggests is, in fact, true: The bottom exact sequence is pure if and only if

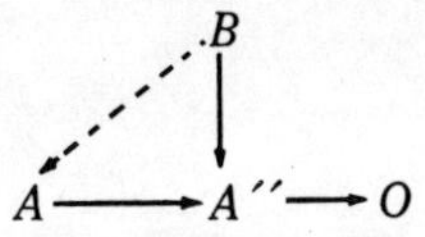

can be completed whenever B is finitely related.

§4. *Particular Rings*

The time has come to give examples. We ask two general questions: If we impose conditions on projective, injective or flat R-modules, how does this affect R? If we impose conditions on R, how does this influence projectives, injectives, and flats? An illustration of the sort of answer we seek is Corollary 3.23: If R is a principal ideal domain, an R-module is injective if and only if it is divisible.

Exercises 4.1. If R is a division ring, then every R-module is injective, projective, and flat.

4.2. If R is a domain not a field, then an R-module A is both injective and projective if and only if $A = 0$.

DEFINITION. A module A is **simple** (or *irreducible*) if $A \neq 0$ and A has no proper submodules; A is **semisimple** if it is a sum of (possibly infinitely many) simple modules.

THEOREM 4.1. *A module A is semisimple if and only if every submodule of A is a summand.*

Proof: Suppose A is semisimple: $A = \Sigma_{k \in K} S_k$, where S_k is simple. Given a subset I of K, let $S_I = \Sigma_{k \in I} S_K$. Now let B be a submodule of A. By Zorn's Lemma, there is a maximal subset I of K with $S_I \cap B = O$. It suffices to show $A = S_I \oplus B$; for this, it suffices to show $S_k \subset S_I \oplus B$ for each $k \in K$. Clearly this inclusion holds for $k \in I$. If $k \notin I$, then maximality of I gives $(S_I + S_k) \cap B \neq O$. If one writes what this means in terms of elements, he sees $(S_I \oplus B) \cap S_k \neq O$. Since S_k is simple, $(S_I \oplus B) \cap S_k = S_k$, i.e., $S_k \subset S_I \oplus B$.

Suppose, conversely, that every submodule of A is a summand.

First, if $B \neq O$ is a submodule of A, we claim B contains a simple submodule. If $b \in B$, $b \neq O$, Zorn's Lemma gives a submodule C of B maximal such that $b \notin C$. Since every submodule of B is a summand of A, hence of B (exercise 2.5), $B = C \oplus D$ for some submodule D. We claim D is simple, for if D' is a proper submodule of D, then $D = D' \oplus D''$ and $B = C \oplus D = C \oplus D' \oplus D''$. Either $C \oplus D'$ or $C \oplus D''$ does not contain b, contradicting the maximality of C. Second, we show A is a sum of simple modules. By Zorn's Lemma, there is a family of simple submodules of A, $\{S_k : k \in K\}$, maximal such that the submodule M they generate is their sum ΣS_k. Now $A = M \oplus B$ for some B. If $B = O$, we are done. If $B \neq O$, then $B = C \oplus S$, where S is simple. The family $\{S, S_k : k \in K\}$ violates the maximality of $\{S_k : k \in K\}$, whence A is semisimple.

COROLLARY 4.2. *Every submodule and every quotient of a semisimple module is semisimple.*

Proof: The property stated in Theorem 4.1 is inherited by submodules.

DEFINITION. A ring R is (left) **semisimple** if, as a left R-module, R is semisimple.

Since the submodules of R are just its left ideals, semisimple rings are sums of (simple) left ideals; moreover, since R has a unit, R is a sum of a finite number of simple left ideals.

Examples 1: Wedderburn's Theorem (Curtis-Reiner, p. 175) says that R is semisimple if and only if

$$R = \sum_{i=1}^{n} M_i ,$$

where M_i is a 2-sided ideal isomorphic (as a ring) to a ring of all square matrices over a division ring D_i. It follows that left semisimple and right semisimple are the same.

2. If G is a (multiplicative) group and F is a field, the **group algebra** FG is an F-algebra whose additive structure is a vector space over F with basis the elements of G and whose multiplication is given by the multiplication of G and the distributive law.

Maschke's Theorem (Curtis-Reiner, p. 41) says that if characteristic F does not divide $|G|$ = order G, (of course, G is finite here) then FG is semisimple. Conversely, if characteristic F does divide $|G|$, then FG is not semisimple.

3. If F is a field of characteristic O, n a positive integer, then $R = F[x]/(x^n - 1)$ is semisimple, for R is just the group algebra of the cyclic group of order n.

4. $R = Z/nZ$ is semisimple when n is square-free.

THEOREM 4.3. *The following are equivalent for a ring R:*

i) *R is left semisimple;*

ii) *R is right semisimple;*

iii) *every left R-module is semisimple;*

iv) *every left R-module is injective;*

v) *every short exact sequence of R-modules is split;*

vi) *every left R-module is projective.*

Proof: We have already observed that i) and ii) are equivalent, by the Wedderburn structure theorem.

i) $\Longrightarrow$ iii) Since R is semisimple as a module, every free module is semisimple. By Corollary 4.2, every quotient of a free module, viz, every module, is semisimple.

iii) $\Longrightarrow$ iv) If A is an R-module, there is an injective module E containing A. Since E is semisimple, A is a summand of E, by Theorem 4.1, and so A is injective.

iv) $\Longrightarrow$ v) This follows from Theorem 3.15.

v) $\Longrightarrow$ vi) If A is an R-module, there is an exact sequence

$$O \to B \to F \to A \to O,$$

where F is free. This sequence splits, so A is a summand of F, hence projective.

vi) $\Longrightarrow$ i) If I is a left ideal of R, then, as every module, R/I is projective. Hence I is a summand of R, by Theorem 3.9. Therefore R is semisimple, by Theorem 4.1.

Because of the equivalence of i) and ii), R is called semisimple (without the adjective "left" or "right") if it satisfies any of the above conditions.

Interesting classes of rings occur when we impose conditions on (1) all modules (or all finitely generated = f.g. modules); (2) all ideals (or all f.g. ideals); (3) R itself considered as a module. One may also ask what the consequences are for R if certain properties are preserved by sums and products. For example, consider injectivity. If all R-modules are injective, we have just seen that R is semisimple. It is an easy exercise (using Theorem 4.1) that if every ideal is injective, R is semisimple also. We defer discussion of rings R for which R itself is injective (for the best results couple this condition with ascending chain conditions). At this point, we only remark that this question leads to quasi-Frobenius and Frobenius rings, examples of which are group algebras FG with no restriction on the characteristic of F.

Let us now ask the same questions for projectivity. If we say every R-module is projective, Theorem 4.3 says R is semisimple. At the other extreme, if we say that R itself is projective, this is no condition at all—R is even free.

DEFINITION. A ring R is **left hereditary** if every left ideal is projective; R is **left semihereditary** if every f.g. left ideal is projective. A hereditary domain is called a **Dedekind ring;** a semihereditary domain is called a **Prüfer ring.**

Examples 1. Any semisimple ring is left hereditary.

2. Any principal ideal domain R is hereditary (for every nonzero ideal is isomorphic to R); it is thus a Dedekind ring.

3. Kaplansky (Nagoya Math. Journal, 1958) gives an example of a right hereditary ring that is not left hereditary. Specifically, if V is a countably infinite dimensional vector space over a field F, C all linear transformations on V with finite dimensional range, B the ring obtained from C by adjoining a unit, then $R = B \otimes_F B$ is right hereditary but not left hereditary.

4. Small (Proc. Nat. Acad. Sci., 1965) gives a simpler example. Let R be the ring of all triangular matrices of the form

$$\begin{pmatrix} a & b \\ 0 & c \end{pmatrix},$$

where $a \in Z$ and $b, c \in Q$. It is easily seen that R is right hereditary but not left hereditary.

THEOREM 4.4. *If R is left hereditary, then every submodule of a free module is isomorphic to a sum of left ideals of R.*

Proof: Let F be a free module with basis $\{x_k\colon k \in K\}$, and suppose the index set K is well-ordered. For each $k \in K$, define

$$F_k = \sum_{i<k} Rx_i \quad \text{and} \quad \bar{F}_k = \sum_{i \leq k} Rx_i \quad .$$

Now let A be a submodule of F. Each $a \in A \cap \bar{F}_k$ has an expression $a = b + rx_k$, where $b \in F_k$, $r \in R$. If $\phi_k : A \cap \bar{F}_k \to R$ is defined by $a \mapsto r$, then there is an exact sequence

$$O \longrightarrow A \cap F_k \longrightarrow A \cap \bar{F}_k \xrightarrow{\phi_k} I_k \longrightarrow O ,$$

where $I_k = \text{im}\, \phi_k$ is an ideal of R. Since I_k is projective, this sequence splits, i.e., $A \cap \bar{F}_k = (A \cap F_k) \oplus C_k$, where $C_k \cong I_k$. We claim that $A = \Sigma_{k \in K} C_k$, which will complete the proof.

(i) *Expression:* We claim A is generated by $\{C_k : k \in K\}$.

Since $F = \cup_{k \in K} \bar{F}_k$, each $a \in A$ (as any element of F) lies in some $\bar{F}_k$. Let $\mu(a)$ be the least k with $a \in \bar{F}_k$. Suppose $A \neq \Sigma_{k \in K} C_k$, and consider $\{\mu(a) : a \in A \text{ and } a \notin \Sigma\, C_k\}$. Let j be the least such, and let $x \in A$ be such that $\mu(x) = j$ and $x \notin \Sigma\, C_k$. Now $\mu(x) = j$ says $x \in A \cap \bar{F}_j$, so that $x = b + c$, $b \in A \cap F_j$ and $c \in C_j$. Therefore $b = x - c \in A$, $b \notin \Sigma\, C_k$ (lest $x \in \Sigma\, C_k$), and $\mu(b) < j$, a contradiction. Hence A is generated by $\{C_k : k \in K\}$.

(ii) *Uniqueness:* Suppose $c_1 + \cdots + c_n = O$, where $c_i \in C_{k_i}$, and $k_1 < k_2 < \cdots < k_n$. Then

$$(c_1 + \cdots + c_{n-1}) + c_n \in (A \cap F_{k_n}) \oplus C_{k_n} .$$

It follows that $c_n \in C_{k_n} \cap (A \cap F_{k_n}) = O$. Induction on n now gives $c_i = O$, all i.

COROLLARY 4.5. *If R is left hereditary, every submodule of a projective module is projective.*

COROLLARY 4.6. *If R is a principal ideal domain, every submodule of a free R-module is free.*

Proof: Every nonzero ideal of R is isomorphic to R.

COROLLARY 4.7. *If R is a principal ideal domain, every projective R-module is free.*

It turns out that Corollary 4.5 (or its dual) characterizes left hereditary rings. We need some general lemmas to prove this.

LEMMA 4.8. *Consider the diagram*

$$\begin{array}{ccccccccc} 0 & \longrightarrow & A' & \xrightarrow{i} & A & \longrightarrow & A'' & \longrightarrow & 0 \\ & & & & \Big\downarrow{\scriptstyle \sigma} & & & & \\ & & & & Q & & & & \end{array}$$

where σ is monic and the top row is exact. Then we may complete this diagram to a commutative diagram with exact rows

$$\begin{array}{ccccccccc} 0 & \longrightarrow & A' & \xrightarrow{i} & A & \longrightarrow & A'' & \longrightarrow & 0 \\ & & \Big\downarrow{\scriptstyle 1_{A'}} & & \Big\downarrow{\scriptstyle \sigma} & & \Big\downarrow & & \\ 0 & \longrightarrow & A' & \longrightarrow & Q & \longrightarrow & Q'' & \longrightarrow & 0 \end{array}$$

Proof: Define $Q'' = \operatorname{coker} \sigma i$, and observe that diagram-chasing gives a map $A'' \to Q''$ making all commute.

Incidentally, we are consistently omitting proofs that involve "diagram-chasing," i.e., proving certain exactness properties of a diagram that follow from assumed commutativity and other exactness properties. The reason for the omission is that the proofs are automatic, but look bad when written out. For example, to prove a map monic, take an element x in the kernel. Now x lies in a vertex of the diagram. There is only one sensible thing to do with x (depending on hypotheses): push it along by an arrow, or lift it backwards because some map is assumed epic. If you can finish the proof of Lemma 4.8, you should be able to chase most diagrams.

LEMMA 4.9. *A module P is projective if and only if every diagram with Q injective*

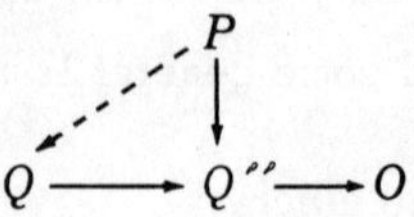

can be completed to a commutative diagram.

Proof: If P is projective, we can complete any such diagram, whether or not Q is injective.

For the converse, suppose we have the diagram

$$\begin{array}{ccccccccc} & & & & & & P & & \\ & & & & & & \downarrow & & \\ 0 & \longrightarrow & A' & \longrightarrow & A & \longrightarrow & A'' & \longrightarrow & 0 \end{array} ;$$

we want a map $P \to A$ making the diagram commute. There is an injective module Q and a monic $\sigma: A \to Q$. By Lemma 4.9, there is a diagram

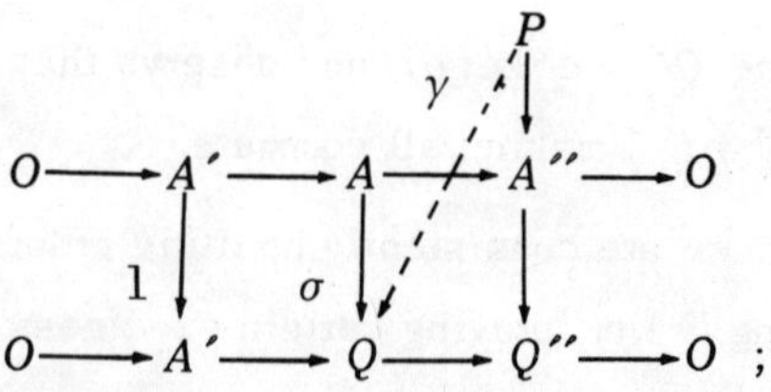

moreover, the dotted arrow exists, by hypothesis. It only remains to prove $\operatorname{im} \gamma \subset \operatorname{im} \sigma$, a routine chase.

We remark that the dual of Lemma 4.9 is true, and its proof is similar. Indeed, one should compare the dual with Theorem 3.19.

THEOREM 4.10. *The following are equivalent for a ring R:*

i) *R is left hereditary;*

ii) *every submodule of a projective module is projective;*

iii) *every quotient of an injective module is injective.*

Proof: i) $\Longrightarrow$ ii) This is Corollary 4.5.

ii) $\Longrightarrow$ i). R itself is projective (even free), so its submodules (the left ideals) are projective, and R is left hereditary.

ii) $\Longrightarrow$ iii). Consider the diagram with exact rows

$$\begin{array}{ccccc} P & \longleftarrow & P' & \longleftarrow & 0 \\ \downarrow & \swarrow & \downarrow & & \\ Q & \longrightarrow & Q'' & \longrightarrow & 0 \end{array} ,$$

where P is projective and Q is injective. By Lemma 4.9, it suffices to find a map $P \to Q$ making the diagram commute. By hypothesis, P' is projective, so there is a map $P' \to Q$ making the diagram commute. By injectivity of Q, we may extend this map from P' to a map $P \to Q$.

iii) $\Longrightarrow$ ii). This is dual to what we have just done.

It follows that if R is not left hereditary, there is some injective Q having a quotient Q'' that is not injective. We know that both Q and Q'' are divisible modules (Theorems 3.20 and 3.21); hence there are divisible modules that are not injective.

EXAMPLES

1. Any left hereditary ring is left semihereditary.
2. R = **valuation ring** (i.e., a domain such that given any two

elements, one divides the other). In such a ring, every f.g. ideal is principal; R is thus a Prüfer ring.

3. Let X be a noncompact Riemann surface, and let R be the ring of all complex-valued functions on X that are analytic. Helmer (*Duke Math. J.*, 1940) proved that R is a domain in which every f.g. ideal is principal, so that R is a Prüfer ring.

4. Chase (*Nagoya Math. J.*, 1961) gives an example of a left semi-hereditary ring that is not right semihereditary.

DEFINITION. A ring R is **regular** (in the sense of von Neumann) if, for each $a \in R$, there is an element $a' \in R$ with $aa'a = a$.

EXAMPLES.

1. R = **Boolean ring** (if $r \in R$, then $r^2 = r$).
2. R = the endomorphism ring of a vector space V (where V is of any dimension, possibly infinite).

Note that if $e = a'a$, then $e^2 = e$, so that regular rings have many idempotents.

THEOREM 4.11. *Every semisimple ring is regular.*

Proof: Suppose a ring $R = I \oplus J$, where I and J are left ideals. Then $1 = x + y$, where $x \in I$ and $y \in J$. Since $x(x + y) = (x + y)x$, we have $xy = yx$; it follows that $xy \in I \cap J = 0$ and $xy = 0$. Furthermore, if $z \in I$, then $z = zx + zy$ so that $zy = z - zx \in I$; on the other hand, $zy \in J$ whence $zy \in I \cap J = 0$ and $zy = 0$.

Now let $a \in R$. Since R is semisimple, there is a left ideal J with $R = Ra \oplus J$. In particular, $1 = ra + y$, where $ra \in Ra$ and $y \in J$. We have just seen that $ay = 0$, so that $a = ara + ay = ara$, and R is regular.

THEOREM 4.12. *Every regular ring R is both left and right semi-hereditary.*

Proof: We prove "left," leaving "right" to the reader. Note that regularity is, by definition, a symmetric condition, so that it is not surprising that the conclusion is not one-sided.

If Ra is a principal left ideal, and if $e = a'a$ (where $aa'a = a$), then $Ra = Re$: $a \in Re$ and $e \in Ra$. Thus, every principal left ideal is generated by an idempotent, hence is a ring summand of R and so is projective. (Alternatively, here is a module theoretic proof: the map $R \to Ra$ defined by $r \mapsto ra'a$ is a projection of R on Ra.)

We shall show that every f.g. left ideal is principal, and this will complete the proof. By induction, it suffices to prove $Ra + Rb$ is principal. There is an idempotent e with $Ra = Re$. Also, $Ra + Rb = Re + Rb(1-e)$ (both a and b are in the right side and both e and $b(1-e)$ are in the left side). There is an idempotent f with $Rb(1-e) = Rf$, namely, $f = cb(1-e)$ for some $c \in R$ (as we saw in the preceding paragraph). It follows that $fe = 0$. We do not know that $ef = 0$, so we adjust f accordingly.

Define $g = (1-e)f$. Then

$$g^2 = (1-e)f(1-e)f = (1-e)(f-fe)f = (1-e)f^2 = (1-e)f = g,$$

$$ge = 0 = eg,$$

and

$$Rg = Rf,$$

so that, finally, $Ra + Rb = Re + Rg$. We claim that $Re + Rg = R(e+g)$. Clearly $Re + Rg \supset R(e+g)$. For the reverse inclusion, note that $re + sg = (re + sg)(e+g) \in R(e+g)$. This completes the proof.

THEOREM 4.13. *If* R *is left semihereditary, then every f.g. submodule of a free module is a sum of a finite number of f.g. left ideals.*

Proof: Let F be a free R-module with basis $\{x_k : k \in K\}$. If A is a f.g. submodule of F, then each generator of A is a finite linear combination of x_k's, so that A is contained in a free summand of F generated by finitely many x_k's. We may, therefore, assume that F is free with basis $\{x_1, \ldots, x_n\}$.

We do an induction on n that A is a sum of f.g. ideals. If $n = 1$, A is isomorphic to a f.g. ideal. Suppose $n > 1$, and define $B = A \cap (Rx_1 \oplus \cdots \oplus Rx_{n-1})$. Each $a \in A$ has a unique expression $a = b + rx_n$, where $b \in B$. If $\phi : A \to R$ is defined by $a \mapsto r$, then there is an exact sequence

$$0 \longrightarrow B \longrightarrow A \xrightarrow{\phi} I \longrightarrow 0$$

where $I = \operatorname{im} \phi$ is a f.g. ideal of R. Since I is projective, this sequence splits, i.e., $A \cong B \oplus I$. Since B is contained in $Rx_1 \oplus \cdots \oplus Rx_{n-1}$, the inductive hypothesis gives B a sum of finitely many f.g. ideals.

COROLLARY 4.14. *R is left semihereditary if and only if every f.g. submodule of a projective R-module is projective.*

Proof: Suppose P is projective and A is a f.g. submodule of P. Since P is a submodule of a free module (even a summand), Theorem 4.13 gives $A \cong \Sigma I_i$, where I_i is isomorphic to a f.g. ideal of R. Since R is left semihereditary, each I_i is projective, and A is projective.

For the converse, R itself is projective, so its f.g. submodules i.e., f.g. left ideals, are projective.

LEMMA 4.15. *Let R be a ring; an R-module A is projective if and only if there are elements $\{a_k : k \in K\}$ of A and homomorphisms $\phi_k : A \to R$ such that*

i) *if $a \in A$, almost all $\phi_k a = 0$;*

ii) *if $a \in A$, $a = \Sigma_k (\phi_k a) a_k$.*

Proof: There is an epimorphism $\psi : F \to A$ where F is a free module with basis $\{e_k : k \in K\}$, and, since A is projective, there is a map $\phi : A \to F$ with $\psi\phi = 1_A$. If $a \in A$, then ϕa (as any element of F) has a unique expression $\phi a = \Sigma\, r_k e_k$, where almost all $r_k = 0$. Define $a_k = \psi e_k$ and define $\phi_k : A \to R$ by $a \mapsto r_k$. Given $a \in A$, almost all $\phi_k a = 0$; moreover,

$$a = \psi\phi a = \psi\, \Sigma\, r_k e_k = \Sigma\, r_k \psi e_k = \Sigma\, (\phi_k a) a_k \quad .$$

Conversely, suppose the elements $\{a_k : k \in K\}$ and the maps $\phi_k : A \to R$ exist. Note that condition (ii) implies that A is generated by the a_k, so there is a free module F with basis $\{e_k : k \in K\}$ and an epimorphism $\psi : F \to A$ with $\psi e_k = a_k$. Define $\phi : A \to F$ by $a \mapsto \Sigma\, (\phi_k a) e_k$, noting that the sum on the right does make sense since almost all $\phi_k a = 0$. Finally, if $a \in A$, then

$$\psi\phi a = \psi\, \Sigma\, (\phi_k a) e_k = \Sigma\, (\phi_k a) a_k = a\,,$$

by condition (ii), so that $\psi\phi = 1_A$, and A is projective.

DEFINITION. Let R be a domain with quotient field Q. An ideal I of R is **invertible** if there are elements $a_1, \ldots, a_n \in I$, $q_1, \ldots, q_n \in Q$ with

i) $q_i I \subset R$, $\quad i = 1, \ldots, n$;

ii) $1 = \Sigma_{i=1}^{n}\, q_i a_i$.

REMARKS. 1. In a domain R, every nonzero principal ideal is invertible.

2. If I is invertible, then $I = Ra_1 + \cdots + Ra_n$, for if $a \in I$, then condition (ii) gives $a = \Sigma\,(q_i a)\,a_i$ and each $q_i a \in R$, by condition (i).

THEOREM 4.16. *If R is a domain, then a nonzero ideal I is projective if and only if it is invertible.*

Proof: If I is projective, there are elements $\{a_k : k \in K\}$ in I and maps $\phi_k : I \to R$ satisfying conditions (i) and (ii) of Lemma 4.15. If $b \in I$, $b \neq O$, define $q_k \in Q$ by

$$q_k = \phi_k(b)/b \ .$$

Note that q_k does not depend on the choice of b: if $b' \in I$, $b' \neq O$, then

$$b'\phi_k(b) = \phi_k(b'b) = \phi_k(bb') = b\phi_k(b')$$

and so

$$\phi_k b/b = \phi_k b'/b' \ .$$

It follows that $q_k I \subset R$ for all k, for $bq_k = b\phi_k(b)/b = \phi_k(b) \in R$. Next, $bq_k = \phi_k(b)$ implies $bq_k = O$ for almost all k, by condition (i), so that if $b \neq O$, almost all $q_k = O$ (all takes place in the field Q). Finally, if $b \in I$,

$$b = \Sigma\,(\phi_k b)\,a_k = \Sigma\,(q_k b)\,a_k = b(\Sigma\, q_k a_k) \ ,$$

so that if $b \neq O$, we may cancel and get $1 = \Sigma\, q_k a_k$. Thus, I is invertible.

Suppose I is invertible, so there are elements a_k, q_k satisfying the definition. Define $\phi_k : I \to R$ by $\phi_k a = q_k a$ (note that im $\phi_k \subset R$ since $q_k I \subset R$). If $a \in I$, then

$$\Sigma\, (\phi_k a)\, a_k = \Sigma\, q_k a a_k = a\, \Sigma\, q_k a_k = a\ .$$

By the lemma, I is projective.

COROLLARY 4.17. *A domain is a Dedekind ring (i.e., is hereditary) if and only if every nonzero ideal is invertible.*

COROLLARY 4.18. *If R is a Dedekind ring, then every ideal of R is f.g.*

Proof: Every invertible ideal is finitely generated.

REMARKS 1. The characterization of Dedekind rings in Corollary 4.17 is often taken as the definition.

2. It can be shown (see Zariski-Samuel, p. 279) that every ideal in a Dedekind ring can be generated by two elements.

3. The ring of integers in a finite extension of the rationals is a Dedekind ring (see Zariski-Samuel, p. 283-284). In particular, there are Dedekind rings that are not principal ideal domains; the standard example is $R = \{a + b\sqrt{-5} : a, b \in Z\}$.

We now generalize Corollary 3.23.

THEOREM 4.19. *A domain R is a Dedekind ring if and only if every divisible module is injective.*

Proof: Suppose R is Dedekind. If D is divisible, we show that, given any nonzero ideal I of R and any map $f: I \to D$, there is an extension of f to R. Since I is invertible, there are elements $q_1, \ldots, q_n$ in Q, $a_1, \ldots, a_n$ in I with $\Sigma\, q_i a_i = 1$ and $q_i I \subset R$.

Since D is divisible, there are elements $d_i \in D$ with $fa_i = a_i d_i$. Now, if $a \in I$,

$$fa = f(\Sigma\, q_i a_i a) = \Sigma\,(q_i a)\, fa_i = \Sigma\,(q_i a a_i)\, d_i = a\, \Sigma\,(q_i a_i)\, d_i\,.$$

Thus, if we set $d = \Sigma\,(q_i a_i)\, d_i$, we have $fa = ad$ for all $a \in I$. Define $g\colon R \to D$ by $g(r) = rd$.

If every divisible module is injective, then Theorem 3.21 implies that every quotient of an injective is injective. By Theorem 4.10, R is hereditary. Since R is a domain, R is Dedekind.

THEOREM 4.20. *If R is a unique factorization domain, then an ideal I is projective if and only if I is principal.*

Proof: If I is principal, then $I = O$ or $I \cong R$ (since R is a domain), so that I is projective.

Conversely, suppose I is projective, and $I \neq O$; by Theorem 4.16, I is invertible. Hence there are elements $q_1, \ldots, q_n \in Q$, $a_1, \ldots, a_n \in I$ with $\Sigma\, q_i a_i = 1$ and $q_i I \subset R$; further, $I = Ra_1 + \cdots + Ra_n$. Write $q_i = b_i/c_i$ and assume (unique factorization) that b_i and c_i have no non-unit factors in common. It follows from $b_i a_j / c_i \in R$ that c_i divides a_j for all i, j. Let $c = \operatorname{lcm}\{c_i\}$. We claim that $I = Rc$. First,

$$c = c\,\Sigma\, b_i a_i / c_i = \Sigma\,(b_i c/c_i)\, a_i \in I\,,$$

for $b_i c/c_i \in R$; therefore $Rc \subset I$. Second, c_i divides a_j, all i, j implies c divides all a_j, i.e., $a_j \in Rc$ for all j.

THEOREM 4.21. *If R is a domain, then every f.g. torsion free module A can be imbedded in a f.g. free module.*

Proof: Let Q be the quotient field of R, and let A have generators $a_1, \ldots, a_n$. It suffices to imbed A in a vector space V over Q, for then A can be imbedded in a finite-dimensional subspace V', say, with basis $\{b_1, \ldots, b_m\}$. Now each $a_i = \Sigma\,(r_{ij}/s_{ij})\,b_j$, where $r_{ij}, s_{ij} \in R$. If $s = \Pi_{i,j}\, s_{ij}$, then $\{s^{-1}b_1, \ldots, s^{-1}b_m\}$ is independent, so that B, the R-submodule of V' generated by this set, is free. Clearly A is imbedded in B.

By Theorem 3.26, A can be imbedded in an injective R-module E. If tE is the torsion submodule of E, then $A \cap tE = O$ (since A is torsion-free), so that A can be imbedded in E/tE. But E/tE is torsion free and divisible (for E is divisible, by Theorem 3.20, and any quotient of a divisible module is divisible). It follows from exercise 3.8 that E/tE is a vector space over Q.

THEOREM 4.22. *A domain R is a Prüfer ring if and only if every f.g. torsion free module A is projective.*

Proof: Suppose R is a Prüfer ring. By Theorem 4.21, A can be imbedded in a free module. By Corollary 4.14, A is projective.

For the converse, note that every ideal in a domain is torsion free.

THEOREM 4.23. *If R is a Prüfer ring, an R-module A is flat if and only if it is torsion free.*

Proof: Suppose A is flat and $O \to K \to F \xrightarrow{\beta} A \to O$ is exact, where F is free. By Theorem 3.37, $K \cap FI = KI$ for every ideal I of R. This holds, in particular, for I principal. Thus, if $ra = O$ and $\beta f = a$, then $rf \in K \cap rF = rK$, i.e., $rf = rk$ for some $k \in K$. But F is torsion free and $r(f-k) = O$, so that $f = k \in K$, $\beta f = O$, and $a = O$. Therefore A is torsion free.

Suppose A is torsion free. By Corollary 3.31, it suffices to show that every f.g. submodule B of A is flat. By Theorem 4.22, B is projective, hence B is flat.

REMARK. It can be shown that the condition: 'Every torsion free module is flat' characterizes Prüfer rings.

The situation when every module is flat is easily treated.

THEOREM 4.24. *A ring R has the property that every right R-module is flat if and only if R is regular.*

Proof: If $a \in R$, then the cyclic right R-module R/aR is flat, so that Theorem 3.37 gives $aRI = aR \cap RI = aR \cap I$, for every left ideal I. In particular, if $I = Ra$, then $a \in aR \cap Ra = aRa$, so there is some $r \in R$ with $a = ara$. Therefore, R is regular.

Assume R is regular, and let B be a right R-module; let $0 \to K \to F \to B \to 0$ be exact, where F is free. By Theorem 3.37, it suffices to prove $KI = K \cap FI$ for every f.g. right ideal I. Now we saw in the proof of Theorem 4.12 that I is principal, so $I = Ra$ for some $a \in R$. We need only show that if $k \in K$ and $k = fa \in Fa$, then $k \in Ka$. But $k = fa = faa'a = ka'a \in Ka$. Therefore, B is flat.

DEFINITION. A ring R is **left noetherian** if every left ideal is f.g.

EXAMPLES

1. Every Dedekind ring.
2. If R is left noetherian, so is $R[x]$ (Hilbert Basis Theorem, see Zariski-Samuel, p. 200).
3. Small's ring of triangular matrices is right noetherian not left noetherian (see page 73).

Exercise 4.3. R is left noetherian if and only if it has the ACC on left ideals, i.e., there are no infinite strictly increasing chains of left ideals.

THEOREM 4.25. *If R is left noetherian, then every submodule of a f.g. left R-module A is also f.g.*

Proof: Suppose A is generated by $x_1, \ldots, x_n$, and B is a submodule of A. We do an induction on n that B is f.g. If $n = 1$, then $A \cong R/I$ for some left ideal I, and so $B \cong J/I$ for some left ideal J of R. Since R is left noetherian, J is f.g. and so B is f.g.

Suppose $n > 1$. If $A' = Rx_1$, there is an exact sequence $0 \to A' \to A \to A/A' \to 0$, and A/A' can be generated by $n-1$ elements. There is another exact sequence:

$$0 \longrightarrow B \cap A' \longrightarrow B \longrightarrow B/(B \cap A') \longrightarrow 0 .$$

By case $n = 1$, $B \cap A'$ is f.g., while induction shows that $B/B \cap A' \cong (B + A')/A' \subset A/A'$ is f.g. It follows that B is f.g.

THEOREM 4.26. *If R is left noetherian, then every f.g. flat R-module* A *is projective.*

Proof: There is an exact sequence $0 \to K \to F \to A \to 0$ where F is f.g. free. By Theorem 4.25, K is f.g., so that A is finitely related. Hence A is projective, by Theorem 3.40.

THEOREM 4.27. *The following are equivalent for a ring R:*

(i) *R is left noetherian;*

(ii) *every direct limit of injective R-modules (directed index set) is injective;*

(iii) *every sum of injectives is injective.*

Proof: (i) $\Longrightarrow$ (ii). Let $\{E_k, \phi_{kj}\}$ be a direct system of injective modules, and consider the diagram

$$\begin{array}{ccccc} & & \varinjlim E_k & & \\ & & \uparrow{\scriptstyle f} & & \\ 0 & \longrightarrow & I & \longrightarrow & R, \end{array}$$

where I is a left ideal of R; we must find a map $g: R \to \varinjlim E_k$ extending f.

Let us agree on notation. Write $I = F/D$, where F is free on $a_1, \ldots, a_s$, and let $\overline{a}_i = a_i + D \in I$. By Theorem 4.25, D can be generated by a finite number of elements, say, $d_1, \ldots, d_t$; also, $d_p = \Sigma\, r_{pi} a_i$, where $r_{pi} \in R$. Finally, let $\varinjlim E_k = (\Sigma\, E_k)/S$. (The reader should glance at the proof of Theorem 2.13 again.)

For each i, $f(\overline{a}_i) = e_{k_i} + S$, where $e_{k_i} \in E_{k_i}$. Choose an index $k \geq k_i$, all i, and define $e^i \in E_k$ by $e^i = \phi_{k_i k} e_{k_i}$; hence $f(\overline{a}_i) = e^i + S$, all i. Since each $d_p \in D$, we have $\Sigma\, r_{pi} e^i + S = 0$ in $\varinjlim E_k$; there is thus an index $q = q_p \geq k$ with $\phi_{kq}(\Sigma\, r_{pi} e^i) = 0$. Choose an index $m \geq q_p$, all p, and define $b^i \in E_m$ by $b^i = \phi_{km} e^i$. The end result of all this is that $f(\overline{a}_i) = b^i + S$ and $\Sigma\, r_{pi} b^i = 0$ for all p.

Define $f': F/D \to E_m$ by $\overline{a}_i \mapsto b^i$, all i. Note that f' is well-fined, for we have been careful that all the relations are sent to 0. Since E_m is injective, there is a map $g': R \to E_m$ extending f'. Finally, define $g: R \to E_m$ extending f'. Finally, define $g: R \to \varinjlim E_k$ by $r \mapsto g'r + S$. Note that g does extend f, for

$$g\overline{a}_i = g'(\overline{a}_i) + S = f'(\overline{a}_i) + S = b^i + S = f\overline{a}_i \quad .$$

Therefore, $\varinjlim E_k$ is injective.

(ii) $\Longrightarrow$ (iii). A finite sum of injectives is always injective, and, by exercise 2.11, an infinite sum is a direct limit (with directed index set) of finite sums.

(iii) $\Longrightarrow$ (i). (This implication is due to H. Bass.) We show that if R is not noetherian, there is an ideal I and a map f from I to a sum of injectives which cannot be extended to R. If R is not noetherian, there is a strictly increasing sequence of ideals $I_1 < I_2 < \cdots$; let $I = \bigcup_{n=1}^{\infty} I_n$. Imbed I/I_n in an injective module E_n. We show that $\Sigma\, E_n$ is not injective.

Let $f_n : I \to I/I_n$ be the natural map. For each $a \in I$, $f_n(a) = O$ for large n, so the map $f\colon I \to \Sigma\, E_n$ given by

$$a \longmapsto (f_1 a, f_2 a, \ldots, f_n a, \ldots) = (f_n a)$$

does, in fact, have its image in $\Sigma\, E_n$. Suppose there is a map $g\colon R \to \Sigma\, E_n$ extending f. If $g(1) = (x_n)$, we claim $x_n \neq O$ for all n. Choose $a \in I$, $a \notin I_n$; then $f_n a \neq O$, so that $ga = fa$ has nonzero n-th coordinate $f_n a$. But $ga = ag(1) = (ax_n)$, so that $f_n a = ax_n$ and $x_n \neq O$. This contradicts the fact that almost all the coordinates of an element in a sum are zero.

We mention that Chase (*Trans. AMS,* 1960) answered the dual question: a ring R is such that every product of projective left R-modules is projective if and only if R has the DCC on principal right ideals and every f.g. right ideal is finitely related. If, in addition, R is commutative, then R is characterized as being *artinian* (i.e., descending chain condition on ideals). In particular, if R is a domain, it is a field.

Exercise 4.4. The following are equivalent for a ring R: (i) R is semisimple; (ii) R is left noetherian and regular; (iii) R is left artinian and regular.

We can now continue the program begun at the beginning of the chapter.

DEFINITION. A ring R is a **quasi-Frobenius ring** if it is both left and right noetherian and if ${}_R R$ is injective.

It can be shown (see, e.g., Jans, *Rings and Homology*, p 78), that the apparent asymmetry of the definition is only virtual; if R is quasi-Frobenius, then R_R is also injective.

Clearly any semisimple ring is quasi-Frobenius.

THEOREM 4.28. *If R is a principal ideal domain and $I = Ra$ is a nonzero ideal, then R/I is quasi-Frobenius.*

Proof: As an R-module, R/I is torsion and cyclic, with generator x, say. A submodule of R/I has the form J/I, hence is cyclic with generator cx, say, where we may assume $bc = a$. Consider the diagram

$$\begin{array}{ccccc} & & R/I & & \\ & & \uparrow{\scriptstyle f} & & \\ O & \longrightarrow & J/I & \longrightarrow & R/I \, . \end{array}$$

Now $f(bx) = sx$ for some $s \in R$. Since $bcx = O$, $O = cf(bx) = csx$, so that $cs \in Ra$; therefore $cs = ra = rbc$ and $s = rb$; thus $f(bx) = sx = rbx$. Define $g: R/I \to R/I$ to be multiplication by r, and note that g is an R/I-map. It follows that R/I is injective. Since R/I is clearly noetherian, it is quasi-Frobenius.

COROLLARY 4.29. (i). *For any $n > 1$, Z/nZ is quasi-Frobenius*; (ii) *If F is a field, then $F[x]/I$ is quasi-Frobenius (if $I \neq O$).*

THEOREM 4.30. *Let R be left and right noetherian; R is quasi-Frobenius if and only if every projective module is injective.*

Proof: If R is quasi-Frobenius, R is injective, and so every free module is injective, by Theorem 4.27. It follows that every projective module is injective.

Conversely, R itself is projective so that ${}_R R$ is injective and R is quasi-Frobenius.

It can be shown (see Curtis and Reiner, p. 402) that every f.g. injective module over a quasi-Frobenius ring is projective.

DEFINITION. Let R be a finite-dimensional algebra over a field F; R is a **Frobenius algebra** if ${}_R R$ and $\text{Hom}_F(R_R, F)$ are isomorphic.

THEOREM 4.31. *If R is a Frobenius algebra, then R is quasi-Frobenius.*

Proof: First, every left or right ideal of R is a vector space over F, so that finite dimensionality of R shows R is both left and right noetherian. Second, the remark after the proof of Theorem 3.25 shows that $\text{Hom}_F(R, F)$ is an injective R-module (for F is a field, hence is an injective F-module). Since ${}_R R \cong \text{Hom}_F(R, F)$, ${}_R R$ is injective and so R is quasi-Frobenius.

LEMMA 4.32. *Let R be a finite-dimensional F-algebra, where F is a field. If there is an F-map $\alpha: R \to F$ whose kernel contains no nonzero left ideals, then R is a Frobenius algebra.*

Proof: Define $\theta: {}_R R \to \text{Hom}_F(R, F)$ by $\theta_r(x) = \alpha(xr)$. It is easy to check that each θ_r is an F-map, for F is assumed to lie in the center of R. Also, there is no difficulty in seeing that θ is an R-map, i.e., $r'\theta_r = \theta_{r'r}$. Next, we claim θ is monic. If $\theta_r = O$, then, for all $x \in R$, $O = \theta_r(x) = \alpha(xr)$, so that $Rr \subset \ker \alpha$. It follows that $Rr = O$ and $r = O$. Finally, assume R has dimension n as a vector space over F; $\mathrm{Hom}_F(R, F)$ is just the dual space of R, so that it, too, has dimension n. Hence θ is an isomorphism, and R is Frobenius.

THEOREM 4.33. *Let F be a field, G a finite group, and $R = FG$. Then R is a Frobenius algebra.*

REMARK: No assumption is made about the characteristic of F.

Proof: By Lemma 4.32, it suffices to exhibit an F-map $\alpha : R \to F$ whose kernel contains no nonzero left ideals. Each $r \in R$ has a unique expression $r = \Sigma_{x \in G}\, m_x x$, where $m_x \in F$. Define $\alpha : R \to F$ by $\alpha(r) = m_1$, the coefficient of 1. If $r = \Sigma\, m_x x$, then $\alpha(x^{-1}r) = m_x$, so that $Rr \subset \ker \alpha$ gives $m_x = 0$, all $x \in G$, and $r = 0$. Therefore $\ker \alpha$ contains no nonzero left ideals of R, and R is a Frobenius algebra.

Exercise 4.5. Let R be quasi-Frobenius. Prove that R is left hereditary if and only if R is semisimple.

We end this chapter with a question and a partial answer. For which rings R is it true that projectives are free? Indeed, we shall be more modest and ask when f.g. projectives are free. Serre raised this question for $R = F[x_1, \ldots, x_n]$, where F is a field. If $n = 1$, then R is a principal ideal domain, and so projectives are free, by Corollary 4.7. If $n = 2$, Seshadri (*Proc. Nat. Acad. Sci.*, 1958) proved projectives are free. If $n \geq 3$, the question is open.

DEFINITION. A ring R is **local** if it has a unique maximal left ideal.

Usually it is assumed that local rings are also commutative and noetherian.

Recall that if R is a ring, its (Jacobson) radical J is the intersection of all maximal left ideals; J is a 2-sided ideal, and if $r \in J$, then $1-r$ is invertible. In particular, if R is local with maximal left ideal M, then $M = J$. Moreover, if $r \notin M$, then r is a unit.

THEOREM 4.34. (Nakayama's Lemma) Let R *be a ring with radical* J. *If* A *is a f.g.* R*-module with* $JA = A$, *then* $A = O$.

Proof: Let $\{a_1, \ldots, a_n\}$ be a minimal generating set of A, i.e., no a_i can be deleted leaving a generating set of A. Since $JA = A$, we have $a_1 = \Sigma_{i=1}^n r_i a_i$, $r_i \in J$. Since $1-r_1$ is invertible, we may express a_1 as a linear combination of the remaining a_i, a contradiction.

THEOREM 4.35. *If* R *is a local ring, every f.g. projective* R*-module* A *is free.*

Proof: Let $\{a_1, \ldots, a_n\}$ be a minimal set of generators of A. Let F be a free R-module with basis $\{x_1, \ldots, x_n\}$, and let $\phi : F \to A$ be given by $x_i \mapsto a_i$. We have an exact sequence

$$O \longrightarrow K \longrightarrow F \xrightarrow{\phi} A \longrightarrow O$$

First, if M is the maximal ideal of R, we claim that $K \subset MF$. For suppose $\Sigma\, r_i x_i$ lies in K but not in MF. Then one of the coefficients, say, r_1, is a unit. But $\Sigma\, r_i a_i = O$, so that

$$a_1 = -r_1^{-1} \left(\sum_{i=2}^{n} r_i a_i \right),$$

contradicting the minimality of $\{a_1, \ldots, a_n\}$. Next, because A is projective, $F = K \oplus A'$, where $A' \cong A$, and so $MF = MK \oplus MA'$. Now $MK \subset K \subset MF$, so that $K = MK \oplus (K \cap MA')$. Since

$K \cap MA' \subset K \cap A' = O$, we have $K = MK$. But K is f.g., being a summand of a f.g. module, so that $K = O$, by Nakayama's Lemma. Therefore A is free.

Kaplansky (*Annals of Math.*, 1958) proved that every projective module over a local ring is free, so the finiteness hypothesis can be removed from Theorem 4.35.

Exercises 4.6. If R is local, prove that any two bases of a free R-module have the same cardinal. (Hint: R/M is a division ring.)

4.7. If R is a local domain, every invertible ideal I is principal.

§5. *Group Extensions*

Let E be a (possibly nonabelian) group, A an abelian normal subgroup, and $G = E/A$. We write E and its subgroup A additively, but we write the quotient group G multiplicatively.

DEFINITION. An **extension** of A by G is an exact sequence

$$O \longrightarrow A \longrightarrow E \longrightarrow G \longrightarrow 1 .$$

REMARK. The definition of extension makes perfectly good sense for any group A, but we assume A abelian throughout.

LEMMA 5.1. *An extension of A by G determines a homomorphism $\theta: G \to \mathrm{Aut}(A)$, where $\mathrm{Aut}(A)$ is the group of all automorphisms of A.*

Proof: If $x \in G$, let $\lambda x \in E$ be a *lifting* of x (i.e., $\lambda x \mapsto x$). Define $\theta_x: A \to A$ by $\theta_x(a) = \lambda x + a - \lambda x$, noting that $\mathrm{im}\, \theta_x \subset A$ since A is normal in E. The function $\theta: G \to \mathrm{Aut}(A)$ sending $x \mapsto \theta_x$ is well-defined; this follows from a short computation using the facts: (i) A is abelian; (ii) if $\lambda' x$ is a second lifting of x, then $\lambda' x - \lambda x \in A$. Finally, θ is a homomorphism, for since the value of θ is independent of choice of liftings, we may choose $\lambda(xy) = \lambda x + \lambda y$.

DEFINITION. If G is a group (perhaps infinite), the **integral group ring** ZG is the ring whose additive group is the free abelian group with basis G, and whose multiplication is determined by the multiplication of G.

A typical element of ZG is $\Sigma_{x \in G}\, m_x x$, where $m_x \in Z$ and almost all $m_x = 0$.

THEOREM 5.2. *If* $0 \to A \to E \to G \to 1$ *is an extension, then* A *is a left* ZG*-module.*

Proof: In brief, G acts on A by conjugation. In detail, if $x \in G$ and $a \in A$, define $xa = \theta_x(a) = \lambda x + a - \lambda x$. Since θ is a homomorphism, $1a = a$ and $(xy)a = x(ya)$. We now define the action of an arbitrary element of ZG by $(\Sigma\, m_x x)a = \Sigma\, m_x(xa)$.

It should now be clear why we write A additively and G multiplicatively.

There is a second way of writing conjugation. Suppose we had defined $\theta'_x(a) = -\lambda x + a + \lambda x$. The reader may check that $\theta'\colon G \to \mathrm{Aut}(A)$ given by $x \mapsto \theta'_x$ is an antihomomorphism: $\theta'(xy) = \theta'(y)\theta'(x)$. This version of conjugation makes A into a right ZG-module, where $ax = \theta'_x(a)$.

Exercises 5.1. If $x \in G$, then $\theta'_x = \theta_{x^{-1}}$.

5.2. Let A be a left ZG-module. Show that A can be made into a right ZG-module by defining $ax = x^{-1}a$. (If G is not abelian, A need not be a 2-sided ZG-module.) Conclude that ZG-modules can, at our pleasure, be considered as either left or right modules. It follows that if A and B are ZG-modules, we can always arrange matters so that $A \otimes_{ZG} B$ makes sense.

5.3. Give an example in which G is abelian (so that ZG is commutative) and a ZG-module A in which $xa \neq ax$ (i.e., the left and right structures do not coincide). [If R is a commutative ring, one usually makes a left R-module A into a right module by defining $ar = ra$; if A happened to be a right R-module, however, there is no reason to expect that these two right actions of R should be the same.]

DEFINITION. A left ZG-module A is **trivial** if $xa = a$ for all $x \in G$ and all $a \in A$.

Exercise 5.4. The following are equivalent for the left ZG-module A, where $0 \to A \to E \to G \to 1$:

i) A is trivial;

ii) A is contained in the center of E;

iii) $\theta \colon G \to \mathrm{Aut}(A)$ is trivial, i.e., $\theta_x = 1$ for all $x \in G$.

Suppose A is a left ZG-module, and $0 \to A \to E \to G \to 1$ is an extension. It may happen that the extension equips A with a left ZG-module structure distinct from its original structure.

DEFINITION. A left ZG-module A **realizes the operators** of an extension $0 \to A \to E \to G \to 1$ if the two left ZG actions on A coincide:

$$xa = \theta_x(a) = \lambda x + a - \lambda x, \quad \text{all } a \in A, \quad x \in G.$$

The *extension problem* is: Given a left ZG-module A, determine all extensions of A by G that realize the operators.

From now on, we shall say **G-module** instead of left ZG-module.

DEFINITION. An extension $0 \to A \to E \xrightarrow{\pi} G \to 1$ is **split** if there is a homomorphism $\lambda \colon G \to E$ with $\pi\lambda = 1_G$; the middle group E is then called a **semidirect product.**

THEOREM 5.3. *An extension $0 \to A \to E \xrightarrow{\pi} G \to 1$ is split if and only if E contains a subgroup $C \cong G$ (not necessarily normal) with $A + C = E$ and $C \cap A = 0$.*

Proof: If the extension splits, then define $C = \operatorname{im} \lambda$. Since $\pi\lambda = 1_G$, λ is monic, so it is an isomorphism between G and C. Now $\operatorname{im}\lambda$ is a transversal of A in E (i.e., a complete set of coset representatives), so that $A + C = E$. Finally, if $a \in A$, then $\pi a = 1$; if $a \in C$, then $a = \lambda x$ for some $x \in G$, and $\pi a = \pi\lambda x = x$. Therefore if $a \in A \cap C$, then $a = \lambda 1 = O$.

Conversely, each $e \in E$ can be written $e = a + c$, $a \in A$, $c \in C$; moreover, this expression is unique because $A \cap C = O$. Note that $\pi | C: C \to G$ is an isomorphism, and define $\lambda: G \to E$ as $(\pi | C)^{-1}$.

DEFINITION. Let $O \to A \to E \xrightarrow{\pi} G \to 1$ be an extension. A **lifting** is a function $\lambda: G \to E$ with $\pi\lambda = 1_G$ and $\lambda 1 = O$.

Exercises 5.5. An extension is split if and only if it has a lifting that is a homomorphism.

5.6. Show there is a split exact sequence

$$O \longrightarrow Z_3 \longrightarrow S_3 \longrightarrow Z_2 \longrightarrow 1,$$

where Z_3 and Z_2 are cyclic of orders 3 and 2 respectively, and S_3 is the symmetric group on 3 letters. Show there is no projection of S_3 onto Z_3 (compare exercise 2.4).

5.7. If an extension $O \to A \to E \to G \to 1$ is split and if A is G-trivial, then E is isomorphic to the direct product of A and G.

THEOREM 5.4. *An extension* $O \to A \to E \to G \to 1$ *is split if and only if*

i) *there is a group* E' *whose elements are all ordered pairs* $(a, x) \in A \times G$ *with addition given by*

$$(a, x) + (a', y) = (a + xa', xy) \ ;$$

ii) *there is a lifting* λ *so that the function* $\phi: E \to E'$ *defined by* $a + \lambda x \longmapsto (a, x)$ *is an isomorphism.*

Proof: Suppose the extension splits, and let $\lambda\colon G \to E$ be a lifting that is a homomorphism. As in Theorem 5.3, every element of E has a unique expression of the form $a + \lambda x$. We compute addition:

$$\begin{aligned}(a+\lambda x) + (a'+\lambda y) &= a + \lambda x + a' - \lambda x + \lambda x + \lambda y \\ &= a + (\lambda x + a' - \lambda x) + \lambda xy \\ &= a + xa' + \lambda xy\ .\end{aligned}$$

It follows that E' is a group and $\phi\colon E \to E'$ is an isomorphism.

Conversely, define a function $\lambda_1\colon G \to E'$ by $x \mapsto (O, x)$, and note λ_1 is a homomorphism. But $\lambda_1 = \phi\lambda$, so that $\lambda = \phi^{-1}\lambda_1$ is a homomorphism and the extension splits.

We now solve the extension problem in the sense that we construct all possible addition tables for "middle groups" E. Suppose an extension $O \to A \to E \to G \to 1$ is given. If $\lambda\colon G \to E$ is a lifting (λ need not be a homomorphism!), then $\operatorname{im}\lambda$ is a transversal of A in E. Since $\lambda 1 = O$, every element of E can be uniquely expressed as $a + \lambda x$. Both $\lambda(xy)$ and $\lambda x + \lambda y$ represent the same coset of A, so we have

$$(*) \qquad \lambda x + \lambda y = [x, y] + \lambda(xy)$$

for some element $[x, y] \in A$.

DEFINITION. The function $[\ \ ]\colon G \times G \to A$ defined by formula (*) is called a **factor set.**

Further, A realizes the operators (there is no other G-structure in sight), so there is a second formula

$$(**) \qquad xa + \lambda x = \lambda x + a\ .$$

THEOREM 5.5. *Let A be a G-module. A function* $[\ \ ]: G \times G \to A$ *is a factor set if and only if*

i) $[x, 1] = O = [1, x]$;

ii) $x[y, z] - [xy, z] + [x, yz] - [x, y] = O$,

for all $x, y, z \in G$.

Proof: Suppose $[\ \ ]$ is a factor set. The first equation follows from (*) and the assumption $\lambda 1 = O$; the second follows from the associative law in E.

For the converse, we must construct an extension of A by G (in which A realizes the operators), and we must then choose a lifting λ so that $[\ \ ]$ satisfies (*). Let E be the set of all ordered pairs $(a, x) \in A \times G$ with addition defined by

$$(a, x) + (a', y) = (a + xa' + [x, y], xy) \quad .$$

(Compare this with the addition in the semidirect product.) Formula (ii) gives associativity, the identity is $(O, 1)$, and

$$-(a, x) = (-x^{-1}a - x^{-1}[x, x^{-1}], x^{-1}) \quad ,$$

so that E is a group. Define $\pi: E \to G$ by $(a, x) \mapsto x$; clearly π is onto with kernel $A = \{(a, 1) : a \in A\}$. Thus $O \to A \to E \xrightarrow{\pi} G \to 1$ is an extension. If we define $\lambda x = (O, x)$, then it is easy to check that A does realize the operators, viz, formula (**) holds, and that $[\ \ ]$ satisfies formula (*).

DEFINITION. $Z^2(G, A)$ is the abelian group of all factor sets under pointwise addition.

Observe that the zero of $Z^2(G, A)$ is the factor set that is identically zero, and it corresponds to the semidirect product.

The definition of factor set arose from considering an extension and it depends on a choice of lifting λ. Changing λ should still give the same extension.

THEOREM 5.6. *Let* $O \to A \to E \to G \to 1$ *be an extension, and let* λ *and* λ': $G \to E$ *be liftings. If* [ ] *and* ( ) *are the corresponding factor sets, then there is a function* $\langle\ \rangle$: $G \to A$ *satisfying*

(i) $\langle 1\rangle = O$;

(ii) $(x, y) - [x, y] = x\langle y\rangle - \langle xy\rangle + \langle x\rangle$

for all $x, y \in G$.

Proof: For any $x \in G$, both λx and $\lambda' x$ lie in the same coset of A, so there is an element $\langle x\rangle \in A$ with

$$\langle x\rangle = \lambda' x - \lambda x \ .$$

Since $\lambda 1 = O = \lambda' 1$, $\langle 1\rangle = O$. We derive the main formula as follows:

$$\begin{aligned} \lambda' x + \lambda' y &= \langle x\rangle + \lambda x + \langle y\rangle + \lambda y \\ &= \langle x\rangle + x\langle y\rangle + \lambda x + \lambda y \\ &\qquad (A \text{ realizes the operators}) \\ &= \langle x\rangle + x\langle y\rangle + [x, y] + \lambda xy \\ &= \{\langle x\rangle + x\langle y\rangle + [x, y] - \langle xy\rangle\} + \lambda' xy \ . \end{aligned}$$

It follows that $(x, y) = \langle x\rangle + x\langle y\rangle + [x, y] - \langle xy\rangle$, and this gives formula (ii) since each term lies in the abelian group A.

DEFINITION. $B^2(G, A)$ is the set of all functions f: $G \times G \to A$ for which there is a function $\langle\ \rangle$: $G \to A$ with $\langle 1\rangle = O$ and

$$f(x, y) = x\langle y\rangle - \langle xy\rangle + \langle x\rangle \ .$$

The elements of $B^2(G, A)$ are called **coboundaries.**

THEOREM 5.7. *$B^2(G, A)$ is a subgroup of $Z^2(G, A)$.*

Proof: A computation is again left to the reader. In particular, one shows that coboundaries satisfy the two properties of Theorem 5.5 and hence are factor sets.

DEFINITION. $e(G, A) = Z^2(G, A)/B^2(G, A)$.

We summarize our results.

COROLLARY 5.8. *Two factor sets arising from an extension via two choices of liftings determine the same element of $e(G, A)$.*

We have been led to the following

DEFINITION. Two extensions $O \to A \to E \to G \to 1$ and $O \to A \to E' \to G \to 1$ are **equivalent** if there are factor sets [ ] and ( ) of each with $[\ ] - (\ ) \in B^2(G, A)$.

A semantic point: equivalence is a relation among extensions and not merely among factor sets. If $[\ ]'$ and $(\ )'$ are two other factor sets of these extensions, then $[\ ]' - (\ )' \in B^2(G, A)$ also.

THEOREM 5.9. *Two extensions of A by G are equivalent if and only if they fit in a commutative diagram*

$$\begin{array}{ccccccccc} O & \longrightarrow & A & \longrightarrow & E & \longrightarrow & G & \longrightarrow & 1 \\ & & \downarrow{\scriptstyle 1_A} & & \downarrow{\scriptstyle \phi} & & \downarrow{\scriptstyle 1_G} & & \\ O & \longrightarrow & A & \longrightarrow & E' & \longrightarrow & G & \longrightarrow & 1 \end{array} .$$

Proof: If the extensions are equivalent, then the construction of Theorem 5.5 allows us to regard E as all $a + \lambda x$ with addition

$$(a+\lambda x) + (a'+\lambda y) = a + xa' + [x, y] + \lambda xy .$$

Similarly, we may regard E' as all $a + \lambda' x$ (denote the factor set here by (x, y)). We are told that

$$[x, y] - (x, y) = x\langle y\rangle - \langle xy\rangle + \langle x\rangle \ .$$

Define $\phi: E \to E'$ by $a + \lambda x \longmapsto a + \langle x\rangle + \lambda' x$. It is obvious the diagram commutes, and a dull computation shows that ϕ is a homomorphism.

Conversely, suppose $\lambda: G \to E$ is a lifting determining a factor set [ ]. By commutativity of the first square, $[x, y] \in A$ gives $\phi[x, y] = [x, y]$, so that $\phi[\ \ ] = [\ \ ]$. By commutativity of the second square, $\phi\lambda: G \to E'$ is also a lifting, and, applying ϕ to the defining equation

$$\lambda x + \lambda y = [x, y] + \lambda xy$$

shows that $\phi[\ \ ]$ is the factor set determined by $\phi\lambda$. If ( ) is any factor set of E', however, we know that ( ) is equivalent to $\phi[\ \ ]$, by Theorem 5.6. Since $\phi[\ \ ] = [\ \ ]$, the two extensions are equivalent.

REMARK. Diagram chasing shows that ϕ must be an isomorphism.

Although the equivalence relation is defined in a natural manner, it can happen that there are inequivalent extensions of A by G with isomorphic middle groups. If p is a prime, we shall see later that $e(Z_p, Z_p) \cong Z_p$, where Z_p is cyclic of order p. If $0 \to Z_p \to E \to Z_p \to 1$ is an extension, then $|E| = p^2$ and E is abelian. We know there are only two abelian groups of order p^2, one cyclic and one elementary, but there are p equivalence classes of extensions. Here is an explicit construction of two such inequivalent extensions. Let p be an odd prime, let A be cyclic of order p with generator a, and let E be cyclic of order p^2 with generator x. Define $i: A \to E$ by $ia = px$; define $j: A \to E$ by $ja = 2px$. Suppose these definitions lead to equivalent extensions, i.e., there is a commutative diagram

$$\begin{array}{ccccccccc}
O & \longrightarrow & A & \xrightarrow{i} & E & \xrightarrow{\pi} & E/iA & \longrightarrow & O \\
 & & 1\downarrow & & \downarrow\beta & & \downarrow 1 & & \\
O & \longrightarrow & A & \xrightarrow{j} & E & \xrightarrow{\pi} & E/jA & \longrightarrow & O
\end{array} .$$

Note that $iA = jA$, so we may take the same map $\pi: E \to E/iA$ in both extensions. Now commutativity of the first square shows that β is multiplication by 2, but this is not compatible with commutativity of the second square.

The Schur-Zassenhaus Lemma (which we will prove eventually) says that if A and G are finite groups with $(|G|, |A|) = 1$, then $e(G, A) = O$. This says that every middle group E of an extension of A by G is a semidirect product. The theorem goes on to say that if C and C' are complements of A in E, then C and C' are conjugate. Let us examine this.

DEFINITION. A **derivation** (or *crossed homomorphism*) is a function $\langle\ \rangle: G \to A$ with

$$\langle xy \rangle = x\langle y \rangle + \langle x \rangle .$$

The set of all derivations, $\mathrm{Der}(G, A)$ is an abelian group under pointwise addition.

If A is a trivial G-module, then $\mathrm{Der}(G, A) = \mathrm{Hom}(G, A)$.

Exercises 5.8. Let A be a semidirect product of A by G, and let $\lambda: G \to E$ be a lifting.

(i) $\lambda x = (\langle x \rangle, x)$ for some $\langle x \rangle \in A$;

(ii) λ is a homomorphism if and only if $\langle\ \rangle$ is a derivation.

5.9 If $\langle\ \rangle$ is a derivation, then $\langle 1 \rangle = O$.

5.10 If $\langle\ \rangle: G \to A$ is a derivation and $h: A \to B$ is a G-map, then $h\langle\ \rangle$ is a derivation.

DEFINITION. An automorphism ϕ of E **stabilizes** an extension $O \to A \to E \to G \to 1$ if the following commutes

$$\begin{array}{ccccccccc} O & \longrightarrow & A & \longrightarrow & E & \longrightarrow & G & \longrightarrow & 1 \\ & & \downarrow 1_A & & \downarrow \phi & & \downarrow 1_G & & \\ O & \longrightarrow & A & \longrightarrow & E & \longrightarrow & G & \longrightarrow & 1 \end{array}.$$

THEOREM 5.10. *Let $O \to A \to E \to G \to 1$ be an extension. There is one-to-one correspondence between stabilizing automorphisms $\phi \in \mathrm{Aut}(E)$ and $\mathrm{Der}(G, A)$.**

Proof: Let $\lambda: G \to E$ be a lifting. Then E consists of all $a + \lambda x$, so that if $\phi \in \mathrm{Aut}(E)$ is stabilizing, then $\phi(a + \lambda x) = a + \langle x\rangle + \lambda x$ for some $\langle\ \rangle: G \to A$. A short computation (using the fact that ϕ is a homomorphism) shows that $\langle\ \rangle$ is a derivation.

We claim that the derivation $\langle\ \rangle$ does not depend on the choice of lifting λ. If $\lambda': G \to E$ is another lifting, then $\phi(a + \lambda' x) = a + [x] + \lambda' x$, and $[x] = \phi\lambda' x - \lambda' x \in A$. Therefore,

$$\begin{aligned} -\langle x\rangle + [x] &= \lambda x - \phi\lambda x + \phi\lambda' x - \lambda' x \\ &= \lambda x + \phi(-\lambda x + \lambda' x) - \lambda' x = O \end{aligned}$$

since $-\lambda x + \lambda' x \in A$, hence is fixed by ϕ. The function $\xi: \phi \mapsto \langle\ \rangle$ is thus well-defined.

To construct a function inverse to ξ, associate to $\langle\ \rangle \in \mathrm{Der}(G, A)$ the map $\phi: E \to E$ defined by $\phi(a + \lambda x) = a + \langle x\rangle + \lambda x$. One checks easily that ϕ is a stabilizing automorphism and that ϕ is independent of the choice of λ.

---

* All stabilizing automorphisms form a subgroup of $\mathrm{Aut}(E)$; the map above is an isomorphism.

THEOREM 5.11. *Let $O \to A \to E \to G \to 1$ be a split extension, and let $\lambda: G \to E$ be a lifting that is a homomorphism. A function $\phi: E \to E$ is an inner stabilizing automorphism if and only if*

$$\phi(a+\lambda x) = a + a_o - xa_o + \lambda x$$

for some fixed $a_o \in A$.

Proof: Suppose ϕ is as displayed. Then ϕ is conjugation by $-a_o$, for

$$-a_o + a + \lambda x + a_o = -a_o + a + xa_o + \lambda x \ .$$

It is clear that ϕ is stabilizing.

Conversely, suppose ϕ is stabilizing and is conjugation by $a_o + \lambda x_o$. Then

$$\begin{aligned} \phi(a+\lambda x) &= a_o + \lambda x_o + a + \lambda x - (a_o + \lambda x_o) \\ &= a_o + x_o a + \lambda x_o + \lambda x - \lambda x_o - a_o \\ &= a_o + x_o a + \lambda x_o x x_o^{-1} - a_o \\ &= a_o + x_o a - x_o x x_o^{-1} a_o + \lambda x_o x x_o^{-1} \\ &= a + \langle x \rangle + \lambda x \ , \end{aligned}$$

since ϕ is stabilizing. It follows that $x_o x x_o^{-1} = x$, so that

$$\phi(a + \lambda x) = a_o + x_o a - xa_o + \lambda x \ .$$

Furthermore, when $x = 1$,

$$a = \phi(a) = a_o + x_o a - a_o \ ,$$

so that $x_o a = a$, and we have the desired formula.

DEFINITION. A **principal derivation** is a function $f: G \to A$ of the form

$$f(x) = xa_o - a_o ,$$

where $a_o \in A$ is fixed.

The set of principal derivations, $\mathrm{P\,Der}(G, A)$, is a subgroup of $\mathrm{Der}(G, A)$.

DEFINITION. $\mathrm{stab}(G, A) = \mathrm{Der}(G, A)/\mathrm{P\,Der}(G, A)$.

THEOREM 5.12. *Let E be a semidirect product of A by G, and let C and C' be complements of A in E. If* $\mathrm{stab}(G, A) = O$, *then C and C' are conjugate.*

Proof: Since E is a semidirect product, there are liftings λ, λ' of G that are homomorphisms and $\mathrm{im}\,\lambda = C$, $\mathrm{im}\,\lambda' = C'$. It follows that the factor sets $[\ \ ]$ and $[\ \ ]'$ they define are identically zero. Theorem 5.6 provides a function $<\ >: G \to A$ with

$$O = [x, y] - [x, y]' = x<y> - <xy> + <x> .$$

It follows that $<\ >$ is a derivation. Since $\mathrm{stab}(G, A) = O$, $<\ >$ is principal, so there is an element $a_o \in A$ with $<x> = xa_o - a_o$. But $<x> = \lambda' x - \lambda x$. Therefore

$$\lambda' x - \lambda x = xa_o - a_o ,$$

and

$$\lambda x = a_o - xa_o + \lambda' x = a_o + \lambda' x - a_o .$$

Since $C = \mathrm{im}\,\lambda$ and $C' = \mathrm{im}\,\lambda'$, a_o conjugates C' into C.

Let us summarize the formulas that have arisen.

Factor set: $x[y, z] - [xy, z] + [x, yz] - [x, y] = O$;

Coboundary: $f(x, y) = x<y> - <xy> + <x>$;

Derivation: $x<y> - <xy> + <x> = O$

Principal derivation: $f(x) = xa - a$.

We can continue this procedure one step further. Factor sets are functions of 2 variables $G \times G \to A$; derivations are functions of 1 variable $G \to A$. Next in line are functions of O variables, i.e., functions from a one point set (denote it [ ]) into A. Now the definition of free G-module says that the group of all functions $G \times \cdots \times G$ (n times) $\to A$ is the same as $\mathrm{Hom}_G(F_n, A)$ where F_n is the free G-module with basis $G \times \cdots \times G$ (n times). A moment's reflection shows that we have been applying the contravariant functor $\mathrm{Hom}_G(\ \ , A)$ to

$$F_3 \xrightarrow{d_3} F_2 \xrightarrow{d_2} F_1 \xrightarrow{d_1} F_o$$

where F_o is the free G-module on the single generator [ ], and

$$d_1[x] = (x-1)[\ \ ];$$

$$d_2[x, y] = x[y] - [xy] + [x];$$

$$d_3[x, y, z] = x[y, z] - [xy, z] + [x, yz] - [x, y] .$$

After all, look at the induced sequence

$$\mathrm{Hom}_G(F_3, A) \xleftarrow{d_3^*} \mathrm{Hom}_G(F_2, A) \xleftarrow{d_2^*} \mathrm{Hom}_G(F_1, A) \xleftarrow{d_1^*} \mathrm{Hom}_G(F_o, A)$$

We see that $\ker d_3^*$ consists of all factor sets (not quite, for we must also take account of the identity $[x, 1] = O = [1, x]$), $\mathrm{im}\, d_2^*$ consists of all coboundaries, $\ker d_2^*$ consists of all derivations, and $\mathrm{im}\, d_1^*$ is all principal derivations. What is $\ker d_1^*$? Note that since $F_o \cong ZG$, we have $\mathrm{Hom}_G(F_0, A) \cong \mathrm{Hom}_G(ZG, A) \cong A$. Thus $\ker d_1^*$ is a submodule of A. Writing these isomorphisms explicitly gives

$$\ker d_1^* = A^G = \{a \in A:\ xa = a, \text{ all } x \in G\} .$$

A^G is called the **fixed points** of A, and it is visibly a G-trivial submodule of A; indeed, it is the maximal G-trivial submodule of A. This is a functor on G-modules: on objects A, its value is A^G; on G-homomorphisms $f\colon A \to B$, its value is the restriction of f to A^G.

THEOREM 5.13. *Let Z be the integers considered as a G-trivial module. Then $t_A\colon \mathrm{Hom}_G(Z, A) \to A^G$ defined by $f \mapsto f(1)$ is an isomorphism.*

Proof: Note that $f(1) \in A^G$, for $xf(1) = f(x \cdot 1) = f(1)$ since f is a G-map and Z is G-trivial. The other verifications are routine.

The reader may check that the sequence $F_3 \to F_2 \to F_1 \to F_o$ is exact. Now $d_1 : F_1 \to F_o = ZG$ is not epic.

DEFINITION. The **augmentation** $\varepsilon : ZG \to Z$ is defined by

$$\Sigma\, m_x x \longmapsto \Sigma\, m_x\ .$$

We let the reader check further that

$$F_1 \xrightarrow{d_1} F_o \xrightarrow{\varepsilon} Z \longrightarrow O$$

is exact.

Exercises 5.11. The augmentation $\varepsilon\colon ZG \to Z$ is a ring homomorphism onto Z. Its kernel is thus a 2-sided ideal in ZG (called the **augmentation ideal**), denoted $\mathcal{G}$.

5.12 If $g\colon A \to B$ is a G-map, the following commutes:

$$\begin{array}{ccc} \mathrm{Hom}_G(Z, A) & \xrightarrow{t_A} & A^G \\ \downarrow g_* & & \downarrow g \\ \mathrm{Hom}_G(Z, B) & \xrightarrow{t_B} & B^G \end{array}$$

It is obviously quite tempting to continue the sequence of free modules past F_3. It is even clear how: with a free resolution of Z. Honest things do appear: the next step $d_4\colon F_4 \to F_3$ yields "obstructions" which tell whether certain types of extensions with non-abelian kernel exist (see MacLane, *Homology*). We will return to this resolution in a later chapter. At this point, the reader has seen that natural group-theoretic problems (surveying extensions and certain conjugacy questions) suggested constructing a free G-resolution of Z, suppressing Z (leaving $\ldots \to F_2 \to F_1 \to F_o \to O$), and applying $\mathrm{Hom}_G(\ ,A)$ to get a complex. The first homology group of this complex is $\mathrm{stab}(G, A)$, the second homology group is almost $e(G, A)$ (for the formula does not take account of the identity $[x, 1] = O = [1, x]$ satisfied by factor sets).

Before leaving groups, we take the topologists' hint of applying $\otimes_{ZG} A$ to the sequence

$$F_2 \xrightarrow{d_2} F_1 \xrightarrow{d_1} F_o \xrightarrow{\varepsilon} Z \longrightarrow O \ .$$

Observe that the free modules in the sequence are left G-modules. As we remarked earlier (exercise 5.2), we may regard them as right G-modules by defining, for $b \in F_n$, $bx = x^{-1}b$, so that the tensor products $F_n \otimes_G A$ make sense. When we applied $\mathrm{Hom}_G(\ ,A)$, we first suppressed Z. Thus, consider

$$F_2 \otimes_G A \xrightarrow{d_2 \otimes 1} F_1 \otimes_G A \xrightarrow{d_1 \otimes 1} F_o \otimes_G A \ ;$$

what are the kernels, images, and quotients?

THEOREM 5.14. $\mathrm{Coker}\ d_1 \otimes 1 \cong Z \otimes_G A$.

Proof: The sequence

$$F_1 \xrightarrow{d_1} F_o \xrightarrow{\varepsilon} Z \longrightarrow O$$

is exact, so that right exactness of $\otimes_G A$ gives exactness of

$$F_1 \otimes A \xrightarrow{d_1 \otimes 1} F_o \otimes A \longrightarrow Z \otimes A \longrightarrow O,$$

whence coker $d_1 \otimes 1 = F_o \otimes A/\text{im}\,(d_1 \otimes 1) \cong Z \otimes_G A.$

THEOREM 5.15. $Z \otimes_G A \cong A/\mathcal{G}A,$ *where* $\mathcal{G}$ *is the augmentation ideal of* $G.$

Proof: Exactness of $O \to \mathcal{G} \to ZG \xrightarrow{\varepsilon} Z \to O$ gives exactness of $\mathcal{G} \otimes_G A \to ZG \otimes_G A \to Z \otimes_G A \to O.$ Now $ZG \otimes_G A \cong A$ and, under this isomorphism, im $\mathcal{G} \otimes A$ goes into $\mathcal{G}A.$

THEOREM 5.16. *As an abelian group,* $\mathcal{G}$ *is free with basis* $\{x - 1:\ x \in G\}.$

Proof: An element $u = \Sigma\, m_x x \in$ ker ε if and only if $\Sigma\, m_x = O.$ Therefore $u = u - (\Sigma\, m_x)1 = \Sigma\, m_x(x - 1),$ so that $\mathcal{G}$ is generated by the $x - 1.$

Suppose $\Sigma\, m_x(x - 1) = O.$ Then $\Sigma\, m_x x - (\Sigma\, m_x)1 = O$ in $ZG,$ which, as an abelian group, is free with basis $G.$ Hence, each $m_x = O.$

DEFINITION. A_G is the maximal quotient module of A that is G-trivial.

COROLLARY 5.17. $A_G \cong A/\mathcal{G}A \cong Z \otimes_G A.$

Proof: It is clear that $A_G = A/S,$ where $S = \{xa - a:\ x \in G,\ a \in A\}$ But $xa - a = (x - 1)\,a,$ and $S = \mathcal{G}A.$

EXAMPLE. Suppose E is a semidirect product of A by G. We may consider G as a subgroup of E, and we inquire about $[G, A]$, the subgroup generated by commutators $a+x-a-x$, $a \in A$, $x \in G$. Now

$$a+x-a-x = a-xa = (1-x)a \ .$$

Therefore $A_G = A/[G, A]$.

Later, we will show that if A = trivial Z, then

$$\ker d_1 \otimes 1/\operatorname{im} d_2 \otimes 1 \ \cong G/G',$$

where G' is the commutator subgroup of G. The next quotient $\ker d_2 \otimes 1/\operatorname{im} d_3 \otimes 1$ also has group-theoretic significance: it is called the *Schur multiplicator*.

THEOREM 5.18. *The function* $\operatorname{Hom}_G(\mathcal{G}, A) \to \operatorname{Der}(G, A)$ *defined by* $f \longmapsto \langle\ \rangle$, *where* $\langle x \rangle = f(x-1)$, *is an isomorphism.*

Proof: First of all, $\langle\ \rangle$ is a derivation. One can prove this directly, or one can observe that $p\colon G \to \mathcal{G}$ defined by $px = x-1$ is a deviation, $\langle\ \rangle = fp$ and use exercise 5.10. Conversely, if $\langle\ \rangle \in \operatorname{Der}(G, A)$, then define a function f by $f(x-1) = \langle x \rangle \in A$. Since $\mathcal{G}$ is a free abelian group with basis $\{x-1\colon x \in G\}$, f can be extended to a Z-homomorphism $\mathcal{G} \to A$. It remains to show that f is a G-map, and this follows from the fact that $\langle\ \rangle$ is a derivation.

This last isomorphism is interesting, for it asserts that the hybrid $\operatorname{Der}(G, A)$ (G is a group, A is a module) can be viewed as an ordinary Hom between two G-modules.

Exercise 5.13. If $g\colon A \to B$ is a G-map, prove the following diagram commutes

$$\begin{array}{ccc} \operatorname{Hom}_G(\mathcal{G}, A) & \longrightarrow & \operatorname{Der}(G, A) \\ \downarrow{\scriptstyle g_*} & & \downarrow{\scriptstyle g_*} \\ \operatorname{Hom}_G(\mathcal{G}, B) & \longrightarrow & \operatorname{Der}(G, B) \end{array}$$

where the horizontal maps are those of Theorem 5.18, and $g_*\colon \operatorname{Der}(G, A) \to \operatorname{Der}(G, B)$ is composition with g.

§6. *Homology*

Throughout this section, module means left R-module, where R is a ring fixed once for all.

DEFINITION. A **complex** $\mathbf{A}$ is a sequence of modules and maps

$$\cdots \longrightarrow A_n \xrightarrow{d_n} A_{n-1} \longrightarrow \cdots$$

$n \in Z$, with $d_n d_{n+1} = O$, all n. The maps d_n are called **differentiations.**

DEFINITION. A **chain map** $f\colon \mathbf{A} \to \mathbf{A}'$ is a family of homomorphisms $f_n\colon A_n \to A'_n$ making the following diagram commute

$$\begin{array}{ccccccc} \cdots \longrightarrow & A_n & \xrightarrow{d} & A_{n-1} & \longrightarrow \cdots \\ & \downarrow f & & \downarrow f & \\ \cdots \longrightarrow & A'_n & \xrightarrow{d'} & A'_{n-1} & \longrightarrow \cdots \end{array}$$

We let the reader check that all complexes and chain maps form a category, Comp, that is pre-additive (the Homs are abelian groups and the distributivity laws hold). We construct homology functors $H_n\colon$ Comp $\to$ abelian groups.

DEFINITION. If $\mathbf{A}$ is a complex, then $d_n d_{n+1} = O$ implies $\operatorname{im} d_{n+1} \subset \ker d_n$. The nth **homology group** $H_n(\mathbf{A})$ is $\ker d_n / \operatorname{im} d_{n+1}$.

The elements of $\ker d_n$ are called n-**cycles**, and the elements of $\operatorname{im} d_{n+1}$ are called n-**boundaries.** One writes $\ker d_n = Z_n(\mathbf{A}) = Z_n$ and $\operatorname{im} d_{n+1} = B_n(\mathbf{A}) = B_n$. Thus

$$H_n(\mathbf{A}) = Z_n(\mathbf{A})/B_n(\mathbf{A}) .$$

To complete the definition of H_n, we define its action on morphisms in Comp. If $f\colon \mathbf{A} \to \mathbf{A}'$ is a chain map, let

$$H_n(f): H_n(\mathbf{A}) \longrightarrow H_n(\mathbf{A}')$$

be given by

$$z_n + B_n \longmapsto f_n z_n + B_n' \,.$$

Usually one writes f_{n*} or even f_* instead of $H_n(f)$.

Observe that $\mathbf{A}$ is exact if and only if $H_n(\mathbf{A}) = O$ for all n. For this reason, exact sequences are also called **acyclic** complexes.

THEOREM 6.1. For each n, H_n : Comp $\to$ *abelian groups is an additive functor.*

Exercises 6.1. A chain map f: $\mathbf{A} \to \mathbf{A}'$ is an isomorphism in Comp if and only if each f_n is an isomorphism.

6.2. Let $\{\mathbf{A}^k\colon k \in K\}$ be a family of complexes, where

$$\mathbf{A}^k = \cdots \longrightarrow A_n^k \xrightarrow{d_n^k} A_{n-1}^k \longrightarrow \cdots$$

Then the *sum* $\Sigma_{k \in K}\ \mathbf{A}^k$ is the complex

$$\cdots \longrightarrow \sum_{k \in K} A_n^k \xrightarrow{\Sigma\, d_n^k} \sum_{k \in K} A_{n-1}^k \longrightarrow \cdots$$

Prove that $H_n(\Sigma_{k \in K}\ \mathbf{A}^k) \cong \Sigma_{k \in K}\ H_n(\mathbf{A}^k)$.

6.3. If f: $\mathbf{A} \to \mathbf{A}'$ is a chain map, define complexes **ker** f, **im** f, and **coker** f in the obvious way: e.g.,

$$\textbf{ker}\, f = \cdots \longrightarrow \ker f_n \xrightarrow{d_n} \ker f_{n-1} \longrightarrow \cdots$$

Prove that $\mathbf{A}/\textbf{ker}\, f \cong \textbf{im}\, f$.

6.4. Define $A' \xrightarrow{f} A \xrightarrow{g} A''$ to be exact if $\ker g = \operatorname{im} f$. Prove that this is exact if and only if

$$A'_n \xrightarrow{f_n} A_n \xrightarrow{g_n} A''_n$$

is exact for each n.

The next natural question is what the homology functors do to short exact sequences of complexes. In general, they are neither left nor right exact (they are "half exact," which we won't define); however, one gets a long exact sequence involving all the H_n's.

THEOREM 6.2. *Let* $O \to A' \xrightarrow{i} A \xrightarrow{p} A'' \to O$ *be an exact sequence of complexes. For each n, there is a homomorphism*

$$\partial_n : H_n(A'') \longrightarrow H_{n-1}(A')$$

defined by

$$z'' + B_n(A'') \longmapsto i^{-1}dp^{-1}z'' + B_{n-1}(A') \quad .$$

Picture:

$$\begin{array}{ccccc} & & A_n & \xrightarrow{p} & A''_n \longrightarrow O \\ & & \downarrow d & & \\ O \longrightarrow & A'_{n-1} & \xrightarrow{i} A_{n-1} & & \end{array}$$

Proof: This is a routine chase, but we supply details because of the importance of the result. We omit some subscripts.

Suppose $z'' \in A''$ with $d''z'' = O$. Since p is epic, we may left z'' to $a \in A_n$, and then push down to $da \in A_{n-1}$. By commutativity,

$$da \in \ker(A_{n-1} \to A''_{n-1}) = \operatorname{im} i \quad ,$$

so that $i^{-1}da$ makes sense, i.e., there is a unique (i is monic) $a' \in A'_{n-1}$ with $ia' = da$. Suppose we had lifted z'' to $a_2 \in A_n$. Then $a - a_2 \in \ker p = \operatorname{im}(A'_n \to A_n)$. This gives an element $a' \in A'_n$ with $a' - a'_2 = d'a' \in B_{n-1}(A')$ (where $ia'_2 = da_2$). Thus we have a well-defined homomorphism

$$Z''_n \longrightarrow A'_{n-1}/B_{n-1}(A') .$$

It is easy to check that this map sends B''_n into O and that $i^{-1}dp^{-1}z''$ is a cycle. Therefore, the formula does give a map

$$H_n(A'') \longrightarrow H_{n-1}(A') .$$

DEFINITION. The homomorphisms ∂_n are called **connecting homomorphisms.**

THEOREM 6.3. (Exact triangle) *If* $O \to A' \xrightarrow{i} A \xrightarrow{p} A'' \to O$ *is an exact sequence of complexes, then there is an exact sequence*

$$\cdots \longrightarrow H_n(A) \xrightarrow{p_*} H_n(A'') \xrightarrow{\partial_n} H_{n-1}(A') \xrightarrow{i_*} H_{n-1}(A) \longrightarrow \cdots$$

Proof: Again the argument is routine, and again we supply details. There are six things to prove; the notation is self-explanatory and we omit subscripts.

1) $\operatorname{im} i_* \subset \ker p_*$.

$$p_* i_* = (pi)_* = O_* = O .$$

2) $\ker p_* \subset \operatorname{im} i_*$.

If $p_*(z + B) = pz + B'' = B''$, then $pz = d''a''$. But $a'' = pa$ for some a, so that $pz = d''pa = pda$, and $p(z - da) = O$. By exactness, there is an a' with $ia' = z - da$. Note that $a' \in Z'$, for $id'a' = dia' = dz - dda = O$ (z is a cycle). Since i is monic,

$d'a' = O$. Therefore, $i_*(a' + B') = ia' + B = z - da + B = z + B$.

3) $\text{im}\, p_* \subset \ker \partial_n$.

$$\partial p_*(z + B) = \partial(pa + B'') = x' + B',$$

where $ix' = dp^{-1}pz = dz = O$ (this computation is correct modulo B'' since ∂ is well-defined). Since i is monic $x' = O$ and $\partial p_* = O$.

4) $\ker \partial_n \subset \text{im}\, p_*$.

If $\partial(z'' + B'') = B'$, then $x' = i^{-1}dp^{-1}z'' \in B'$. Hence $x' = d'a'$. Now $ix' = id'a' = dia' = dp^{-1}z''$. so that $d(p^{-1}z'' - ia') = O$ and $p^{-1}z'' - ia' \in Z$. Therefore

$$p_*(p^{-1}z'' - ia' + B) = pp^{-1}z'' - pia' + B'' = z'' + B''.$$

5) $\text{im}\, \partial_n \subset \ker i_*$.

$i_*\partial_n(z'' + B'') = ix' + B$, where $ix' = dp^{-1}z'' \in B$.

6) $\ker i_* \subset \text{im}\, \partial_n$.

Suppose $i_*(z' + B') = iz' + B = B$, so that $iz' = da$. Then $d''pa = pda = piz' = O$ whence $pa \in Z''$. But $\partial(pa + B'') = x' + B'$, where $ix' = dp^{-1}pa = da = iz'$. Since i is monic, $x' = z'$ and $\partial(pa + B'') = z' + B'$.

Theorem 6.3 is called the exact triangle because of the mnemonic diagram

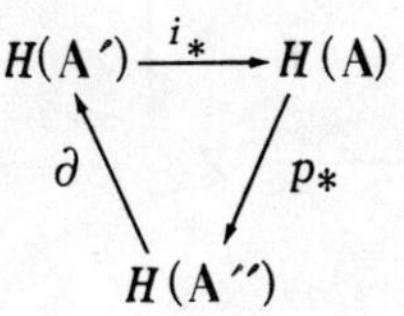

Exercises 6.5 (Serpent lemma) Consider the commutative diagram with exact rows

$$\begin{array}{ccccccc} & & A & \longrightarrow & B & \xrightarrow{p} & C & \longrightarrow & O \\ & & \downarrow{\scriptstyle\alpha} & & \downarrow{\scriptstyle\beta} & & \downarrow{\scriptstyle\gamma} & & \\ O & \longrightarrow & A' & \xrightarrow{i} & B' & \longrightarrow & C' & & \end{array}$$

Prove that D: $\ker \gamma \to \operatorname{coker} \alpha$ defined by $c \mapsto i^{-1}\beta p^{-1}c + \operatorname{im} \alpha$ is a homomorphism. Moreover, there is an exact sequence

$$\ker \alpha \to \ker \beta \to \ker \gamma \xrightarrow{D} \operatorname{coker} \alpha \to \operatorname{coker} \beta \to \operatorname{coker} \gamma$$

6.6 If, in the diagram of the serpent lemma, we further assume $A \to B$ monic and $B' \to C'$ epic, then we have $\ker \alpha \to \ker \beta$ monic and $\operatorname{coker} \beta \to \operatorname{coker} \gamma$ epic.

6.7 Here is a second proof of the exact triangle. Given the commutative diagram with exact rows

$$\begin{array}{ccccccccc} O & \longrightarrow & A'_n & \longrightarrow & A_n & \longrightarrow & A''_n & \longrightarrow & O \\ & & \downarrow{\scriptstyle d'_n} & & \downarrow{\scriptstyle d_n} & & \downarrow{\scriptstyle d''_n} & & \\ O & \longrightarrow & A'_{n-1} & \longrightarrow & A_{n-1} & \longrightarrow & A''_{n-1} & \longrightarrow & O, \end{array}$$

there is another commutative diagram with exact rows

$$\begin{array}{ccccccccc} & & A'_n/\operatorname{im} d'_{n+1} & \longrightarrow & A_n/\operatorname{im} d_{n+1} & \longrightarrow & A''_n/\operatorname{im} d''_{n+1} & \longrightarrow & O \\ & & \downarrow{\scriptstyle d'} & & \downarrow{\scriptstyle d} & & \downarrow{\scriptstyle d''} & & \\ O & \longrightarrow & Z'_{n-1} & \longrightarrow & Z_{n-1} & \longrightarrow & Z''_{n-1} & & \end{array}$$

(Exactness of the rows is exercise 6.6; the downward maps are, e.g., $a'_n + \operatorname{im} d'_{n+1} \mapsto d'_n a'_n$.) Apply the serpent lemma to the second diagram.

THEOREM 6.4. (Naturality of ∂). *Consider the commutative diagram with exact rows of complexes*

$$\begin{array}{ccccccccc} 0 & \longrightarrow & A' & \xrightarrow{i} & A & \xrightarrow{p} & A'' & \longrightarrow & 0 \\ & & \downarrow{\scriptstyle \alpha'} & & \downarrow{\scriptstyle \alpha} & & \downarrow{\scriptstyle \alpha''} & & \\ 0 & \longrightarrow & C' & \xrightarrow{j} & C & \xrightarrow{q} & C'' & \longrightarrow & 0 \end{array}$$

Then there is a commutative diagram with exact rows

$$\begin{array}{ccccccccccc} \cdots \to & H_n(A') & \xrightarrow{i_*} & H_n(A) & \xrightarrow{p_*} & H_n(A'') & \xrightarrow{\partial} & H_{n-1}(A') & \longrightarrow \cdots \\ & \downarrow{\scriptstyle \alpha'_*} & & \downarrow{\scriptstyle \alpha_*} & & \downarrow{\scriptstyle \alpha''_*} & & \downarrow{\scriptstyle \alpha'_*} & \\ \cdots \to & H_n(C') & \xrightarrow{j_*} & H_n(C) & \xrightarrow{q_*} & H_n(C'') & \xrightarrow{\partial} & H_{n-1}(C') & \longrightarrow \cdots \end{array}$$

Proof: The first two squares commute because H_n is a functor. A routine but long chase gives commutativity of the square involving ∂.

Given a functor T, we wish to construct a sequence of new functors (called derived functors) as suggested by our study of group extensions. Here is a rough sketch of how to evaluate these functors on a module A: choose a resolution of A, apply T to it, and then take homology of the resulting complex. It is clear from this description that we are obliged to show independence of the choice of resolution. We prepare the way, again taking a hint from the topologists.

DEFINITION. Let f: $A \to A'$ be a chain map; f is **nullhomotopic** if there are maps $s_n: A_n \to A'_{n+1}$ with

$$f_n = d'_{n+1}s_n + s_{n-1}d_n, \qquad \text{all } n.$$

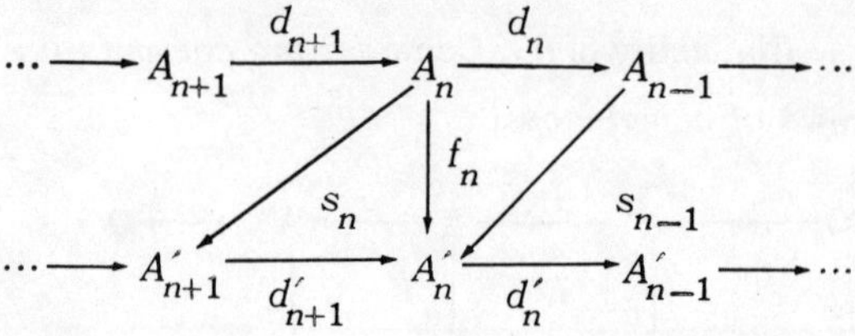

If f and g: $\mathbf{A} \to \mathbf{A}'$ are chain maps, f is **homotopic** to g if $f - g$ is nullhomotopic.

Homotopy is an equivalence relation on the group of all chain maps $\mathbf{A} \to \mathbf{A}'$.

THEOREM 6.5. *Let f, g: $\mathbf{A} \to \mathbf{A}'$ be chain maps. If f and g are homotopic, then, for all n,*

$$f_{*n} = g_{*n} : H_n(\mathbf{A}) \to H_n(\mathbf{A}') \quad .$$

Proof: We omit subscripts. If z is an n-cycle, then

$$f(z) - g(z) = d's(z) + sd(z) \quad .$$

Since $d(z) = O$, $f(z) - g(z) \in B(\mathbf{A}')$ and $f_* = g_*$.

Exercise 6.8. A complex $\mathbf{A}$ has a **contracting homotopy** if the iden- identity map $1_{\mathbf{A}}$ is nullhomotopic. Prove that a complex with a contracting homotopy is an exact sequence.

Notation. Let $\cdots \to X_2 \to X_1 \to X_o \to A \to O$ be a complex. Then $\mathbf{X}_A$ is a **deleted complex** and is defined as the complex

$$\cdots \longrightarrow X_2 \longrightarrow X_1 \longrightarrow X_o \longrightarrow O \quad .$$

Similarly, we define the deleted complex of a complex

$$O \longrightarrow B \longrightarrow Y_o \longrightarrow Y_1 \longrightarrow \cdots$$

by suppressing B; it shall be denoted $\mathbf{Y}_B$.

THEOREM 6.6 (Comparison Theorem). *Consider the diagram*

$$\begin{array}{ccccccccc} \cdots \longrightarrow & X_2 & \xrightarrow{d_2} & X_1 & \xrightarrow{d_1} & X_o & \xrightarrow{\varepsilon} & A & \longrightarrow O \\ &&&&&&& \Big\downarrow f & \\ \cdots \longrightarrow & X'_2 & \xrightarrow{d'_2} & X'_1 & \xrightarrow{d'_1} & X'_o & \xrightarrow{\varepsilon'} & A' & \longrightarrow O\ , \end{array}$$

where the rows are complexes. If each X_n in the top row is projective and if the bottom row is exact, then there is a chain map $\overline{f}: \mathbf{X}_A \to \mathbf{X}'_A$, with

$$\begin{array}{ccc} X_o & \xrightarrow{\varepsilon} & A \\ {\scriptstyle \overline{f}_o}\Big\downarrow & & \Big\downarrow{\scriptstyle f} \\ X'_o & \xrightarrow{\varepsilon'} & A' \end{array}$$

commuting; moreover, any two such chain maps are homotopic.

Proof: (i) The existence of $\overline{f}$.

The proof is an induction on n. If $n = O$, we have the diagram

$$\begin{array}{ccccc} & & X_o & & \\ & & \Big\downarrow{\scriptstyle f\varepsilon} & & \\ X'_o & \xrightarrow{\varepsilon'} & A' & \longrightarrow & O\ . \end{array}$$

Since ε' is epic and X_o is projective, there is a map $\overline{f}_o : X_o \to X'_o$ with $\varepsilon'\overline{f}_o = f\varepsilon$.

Fot the inductive step, consider the diagram

$$\begin{array}{ccccc} X_{n+1} & \xrightarrow{d_{n+1}} & X_n & \xrightarrow{d_n} & X_{n-1} \\ & & \Big\downarrow{\scriptstyle \overline{f}_n} & & \Big\downarrow{\scriptstyle \overline{f}_{n-1}} \\ X'_{n+1} & \xrightarrow{d'_{n+1}} & X'_n & \xrightarrow{d'_n} & X'_{n-1} \end{array}$$

If we show that $\operatorname{im} \bar{f}_n d_{n+1} \subset \operatorname{im} d'_{n+1}$, *then* we will have the diagram

$$\begin{array}{ccccc} & & X_{n+1} & & \\ & & \downarrow{\scriptstyle \bar{f}_n d_{n+1}} & & \\ X'_{n+1} & \xrightarrow{d'_{n+1}} & \operatorname{im} d'_{n+1} & \longrightarrow & O \end{array}$$

and projectivity of X_{n+1} will provide a map $\bar{f}_{n+1}: X_{n+1} \to X'_{n+1}$ with $d'_{n+1}\bar{f}_{n+1} = \bar{f}_n d_{n+1}$. To check the inclusion, note that exactness of the bottom row gives $\operatorname{im} d'_{n+1} = \ker d'_n$, so it suffices to prove $d'_n \bar{f}_n d_{n+1} = O$; but $d'_n \bar{f}_n d_{n+1} = \bar{f}_{n-1} d_n d_{n+1} = O$.

(ii) If $\bar{f}$ and h are two such chain maps, they are homotopic.

We construct a homotopy s by induction. We begin by defining $s_{-1}: X_{-1} \to X'_o$ as the zero map (not much choice here since $X_{-1} = O$). For the inductive step, we will show that

$$\operatorname{im}(h_{n+1} - \bar{f}_{n+1} - s_n d_{n+1}) \subset \operatorname{im} d'_{n+2},$$

so that we will have the picture

$$\begin{array}{ccccc} & & X_{n+1} & & \\ & & \downarrow{\scriptstyle h_{n+1} - \bar{f}_{n+1} - s_n d_{n+1}} & & \\ X'_{n+2} & \xrightarrow{d'_{n+2}} & \operatorname{im} d'_{n+2} & \longrightarrow & O \end{array}$$

Projectivity of X_{n+1} will then give a map $s_{n+1}: X_{n+1} \to X'_{n+2}$ satisfying the desired equation. To check the inclusion, note that exactness of the bottom row gives $\operatorname{im} d'_{n+2} = \ker d'_{n+1}$, so that it suffices to show $d'_{n+1}(h_{n+1} - \bar{f}_{n+1} - s_n d_{n+1}) = O$. This last term is

$$d'_{n+1}(h_{n+1} - \bar{f}_{n+1}) - d'_{n+1} s_n d_{n+1}$$

$$= d'_{n+1}(h_{n+1} - \overline{f}_{n+1}) - (h_n - \overline{f}_n - s_{n-1}d_n)d_{n+1}$$

$$= d'_{n+1}(h_{n+1} - \overline{f}_{n+1}) - (h_n - \overline{f}_n)d_{n+1} ,$$

and this is zero since h and $\overline{f}$ are chain maps.

The dual of Theorem 6.6 is also true and the proof is similar; now the sequences go off to the right ($O \to A \to E_o \to E_1 \to \cdots$), the top row is assumed exact, and each term in the bottom row (save the first) is assumed injective.

DEFINITION. If $\overline{f} : \mathbf{X}_A \to \mathbf{X}'_{A'}$ is a chain map for which $f\varepsilon = \varepsilon' f_o$, we say $\overline{f}$ is **over** f.

Given a functor T, we now describe its **left derived functors** L_nT. For each module A, choose, once for all, a projective resolution of A, and let $\mathbf{P}_A$ be the corresponding deleted complex. Next, apply T to $\mathbf{P}_A$ to get the complex

$$\cdots \longrightarrow TP_2 \xrightarrow{Td_2} TP_1 \xrightarrow{Td_1} TP_o \longrightarrow O .$$

Finally, take homology.

DEFINITION. For each module A, $(L_nT)A = H_n(T\mathbf{P}_A) =$ $\ker Td_n / \operatorname{im} Td_{n+1}$.

To complete the definition of L_nT, we must describe its action on $f: A \to B$. By the comparison theorem 6.6, there is a chain map $\overline{f} : \mathbf{P}_A \to \mathbf{P}_B$ over f. Define

$$(L_nT)f: L_nTA \to L_nTB$$

by

$$(L_nT)f = H_n(T\overline{f}),$$

i.e., if $z_n \in \ker Td_n$, then

$$z_n + \operatorname{im} Td_{n+1} \longmapsto (T\bar{f}_n) z_n + \operatorname{im} Td'_{n+1} .$$

Now the comparison theorem tells us that $L_n T(f)$ is well-defined, for if $h: \mathbf{P}_A \to \mathbf{P}_B$ is a second chain map over f, then $\bar{f}$ and h are homotopic, hence $T\bar{f}$ and Th are homotopic, and so they induce the same maps on homology groups (Theorem 6.5).

Exercise 6.9. Each $L_n T$ is an additive functor.

DEFINITION. Let $T = \otimes_R B$; then $\operatorname{Tor}_n^R(\ ,B) = L_n T$. In particular,

$$\operatorname{Tor}_n^R(A, B) = \ker d_n \otimes 1/\operatorname{im} d_{n+1} \otimes 1 ,$$

where

$$\cdots \longrightarrow P_2 \xrightarrow{d_2} P_1 \xrightarrow{d_1} P_o \longrightarrow A \longrightarrow O$$

is the projective resolution of A that we chose once for all.

The reader should be annoyed that our definition of the functors $L_n T$ depends on a choice of resolution. Had we begun with other choices $\ldots \to \hat{P}_2 \to \hat{P}_1 \to \hat{P}_o \to A \to O$ of projective resolutions, we would have obtained functors (we temporarily denote) $\hat{L}_n T$. Our next project is to show that $L_n T$ and $\hat{L}_n T$ are essentially the same.

DEFINITION. Let $F, F': \mathcal{C} \to \mathcal{C}'$ be functors. Then F and F' are **naturally equivalent** if, for each $A \in \operatorname{obj} \mathcal{C}$, there is an isomorphism $t_A: FA \to F'A$ so that the following diagrams commute for all $g: A \to B$:

$$\begin{array}{ccc} FA & \xrightarrow{t_A} & F'A \\ {\scriptstyle Fg}\downarrow & & \downarrow{\scriptstyle F'g} \\ FB & \xrightarrow{t_B} & F'B \end{array}$$

There is an analogous definition if both F and F' are contravariant.

If we drop the requirement that each t_A be an isomorphism, then $t = \{t_A\colon\ A \in \text{obj}\ \mathcal{C}\}$ is called a **natural transformation,** and one writes $t\colon\ F \to F'$.

We have already seen many examples.

1. $\text{Hom}_R(R,\ \ )$ is naturally equivalent to the identity functor I_R on ${}_R\mathfrak{M}$ (exercises 2.15 and 2.16);
2. $R \otimes_R$ is naturally equivalent to I_R (exercises 2.17 and 2.18).
3. If G is a group and Z is the integers considered as a trivial G-module, then $\text{Hom}_G(Z,\ \ )$ is naturally equivalent to the fixed point functor (Theorem 5.13 and exercise 5.12).
4. If $\mathcal{G}$ is the augmentation ideal of G, then $\text{Hom}_G(\mathcal{G},\ \ )$ and Der $(G,\ \ )$ are naturally equivalent (Theorem 5.18 and exercise 5.13).

Exercises 6.10. Let $\mathcal{C}$ be the category of finite-dimensional vector spaces over a field k. If $*$ is the functor $\text{Hom}_k(\ \ , k)$, show that $**$ is naturally equivalent to the identity functor.

6.11. Let $\mathcal{C}$ be the category of finite abelian groups. If $*$ is the functor $\text{Hom}_Z(\ \ , Q/Z)$, Q = rationals, then $**$ is naturally equivalent to the identity functor. (This is a special case of Pontrjagin duality, where $\mathcal{C}$ is the category of locally compact abelian topological groups and continuous homomorphisms, and * is Hom(, R/Z) suitably topologized $-R$ = reals.)

6.12. Let G, F be an adjoint pair of functors, where $F: \mathfrak{A} \to \mathfrak{B}$ and $G: \mathfrak{B} \to \mathfrak{A}$. Prove there are natural transformations $t: 1_{\mathfrak{A}} \to GF$ and $s: FG \to 1_{\mathfrak{B}}$ (where $1_{\mathfrak{A}}$ is the identity functor on $\mathfrak{A}$). (Hint: For any $A \in \operatorname{obj} \mathfrak{A}$, we have $\operatorname{Hom}(FA, FA) \cong \operatorname{Hom}(A, GFA)$.)

Ordinarily one tends to identify naturally equivalent functors. In particular, if F is naturally equivalent to the identity functor I, one tends to identify each $A = IA$ with FA (e.g., an R-module A is identified with $R \otimes_R A$). The reason is that, whenever we have A in the midst of a commutative diagram, we may pluck it out, replace it by FA, adjust maps by isomorphisms, and, in the end, have a diagram with precisely the same commutativity and exactness properties enjoyed by the original diagram.

THEOREM 6.7. *For any functor T, the functors L_nT and $\hat{L}_nT$ are naturally equivalent. In particular, for each A, $L_nT(A) \cong \hat{L}_nT(A)$ and so these groups are independent of the choice of projective resolution of A.*

Proof: Consider the diagram

$$\begin{array}{ccccccccc}
\cdots \longrightarrow & P_2 & \longrightarrow & P_1 & \longrightarrow & P_o & \longrightarrow & A & \longrightarrow O \\
 & & & & & & & \big\downarrow 1_A & \\
\cdots \longrightarrow & \hat{P}_2 & \longrightarrow & \hat{P}_1 & \longrightarrow & \hat{P}_o & \longrightarrow & A & \longrightarrow O
\end{array}$$

where the top row is the chosen projective resolution of A used to define L_nT, the bottom row that used to define $\hat{L}_nT$. By the comparison theorem, there is a chain map $t: \mathbf{P}_A \to \hat{\mathbf{P}}_A$ over 1_A (unique to homotopy), and, applying T gives a chain map

$Tt: T\mathbf{P}_A \to T\hat{\mathbf{P}}_A$ over 1_{TA}; this latter induces maps

$$Tt_*: L_n T(A) \to \hat{L}_n T(A) .$$

First, we claim these maps are isomorphisms. To get the inverse, turn the diagram above upside down, so that the hatted sequence is now on top. The comparison theorem gives a chain map $s: \hat{\mathbf{P}}_A \to \mathbf{P}_A$. The composite $st: \mathbf{P}_A \to \mathbf{P}_A$ is thus a chain map over 1_A. Since $1_{\mathbf{P}}: \mathbf{P}_A \to \mathbf{P}_A$ is also a chain map over 1_A, st and $1_{\mathbf{P}}$ are homotopic and $s_*t_* = 1$. Similarly $t_*s_* = 1$, so that t_* is an isomorphism. Therefore Tt_* is an isomorphism.

Second, we show the following diagram commutes for $f: A \to B$:

$$\begin{array}{ccc} L_n T(A) & \xrightarrow{Tt_*} & \hat{L}_n T(A) \\ \downarrow & & \downarrow \\ L_n T(B) & \xrightarrow{Tt_*} & \hat{L}_n T(B) \end{array} .$$

To evaluate in the clockwise direction, consider the picture

$$\begin{array}{ccccccccc} \cdots & \longrightarrow & P_1 & \longrightarrow & P_o & \longrightarrow & A & \longrightarrow & O \\ & & & & & & \downarrow 1_A & & \\ \cdots & \longrightarrow & \hat{P}_1 & \longrightarrow & \hat{P}_o & \longrightarrow & A & & \\ & & & & & & \downarrow f & & \\ \cdots & \longrightarrow & \hat{Q}_1 & \longrightarrow & \hat{Q}_o & \longrightarrow & B & & \end{array}$$

Applying the comparison theorem gives a chain map $\mathbf{P}_A \to \hat{\mathbf{Q}}_B$ over $f 1_A = f$. Going counter clockwise also gives a chain map $\mathbf{P}_A \to \hat{\mathbf{Q}}_B$ over $1_B f = f$. Applying T to these gives chain maps $T\mathbf{P}_A \to T\hat{\mathbf{Q}}_B$ over Tf which are homotopic (since the original chain maps are); the maps they induce are thus equal.

COROLLARY 6.8. *The definition of* $\mathrm{Tor}_n^R(A, B)$ *is independent of the choice of projective resolution of A.*

We still have to grapple with the question: Suppose we begin with the functor $T = A \otimes_R$ and carry out the construction of its left derived functors; if we call the resulting functors $\mathrm{Tor}_n^R(A, \ )$, is $\mathrm{Tor}_n^R(A, B) \cong \mathrm{Tor}_n^R(A, B)$? There is no misprint. We ask whether the value of $\mathrm{Tor}_n^R(A, \ )$ on B is the same as the value of $\mathrm{Tor}_n^R(\ , B)$ on A. Before we worry about this, let us construct right derived functors. The recipe is almost the same as for left derived functors, so that our sketch of the construction should serve as a review.

There are two definitions, depending on the variance of T (there is a similar dichotomy in the construction of $L_n T$, but we will not worry about $L_n T$ when T is contravariant).

For each module A, choose, once for all, an injective resolution

$$0 \longrightarrow A \longrightarrow E_0 \longrightarrow E_1 \longrightarrow \cdots$$

We need a consistent notation. In the very definition of complex, the indices go down as we go to the right ($A_n \to A_{n-1}$). To avoid troubles, write the injective resolution

$$0 \longrightarrow A \longrightarrow E_0 \xrightarrow{d_0} E_{-1} \xrightarrow{d_{-1}} E_{-2} \longrightarrow \cdots .$$

Let $\mathbf{E}_A$ be the associated deleted resolution.

DEFINITION. It T is covariant, define the **right derived functors** $R^n T$ on modules A by

$$R^n T(A) = H_{-n}(T\mathbf{E}_A) = \ker Td_{-n} / \mathrm{im}\ Td_{-n+1} .$$

The definition of $R^nT(f)$, for $f\colon A \to B$, works just as before. The dual of the comparison theorem asserts the existence of a chain map $\overline{f}\colon \mathbf{E}_A \to \mathbf{E}_B$ that is unique up to homotopy, so there are unique maps $H_{-n}(T\mathbf{E}_A) \to H_{-n}(T\mathbf{E}_B)$ defined for each n induced by $T\overline{f}$.

Exercises. *6.13* If T is covariant, each R^nT is an additive functor.

6.14. If T is covariant, and if $\hat{R}^nT$ is defined as R^nT via different choices of injective resolutions, then R^nT and $\hat{R}^nT$ are naturally equivalent.

DEFINITION. If $T = \operatorname{Hom}_R(C, \ )$, then $R^nT = \operatorname{Ext}^n_R(C, \ )$. In particular,

$$\operatorname{Ext}^n_R(C, A) = \frac{\ker \operatorname{Hom}(C, d_n)}{\operatorname{im} \operatorname{Hom}(C, d_{n-1})},$$

where $O \to A \to E_o \xrightarrow{d_o} E_1 \xrightarrow{d_1} E_2 \to \cdots$ is the chosen injective resolution of A.

COROLLARY 6.9. The definition of $\operatorname{Ext}^n_R(A, B)$ is independent of the choice of injective resolution of B.

Finally, we consider T contravariant. Start with projective resolutions

$$\cdots \longrightarrow P_2 \xrightarrow{d_2} P_1 \xrightarrow{d_1} P_o \longrightarrow A \longrightarrow O$$

and apply T to the deleted resolution $\mathbf{P}_A$. We obtain

$$O \longrightarrow TP_o \xrightarrow{Td_1} TP_1 \xrightarrow{Td_2} TP_2 \longrightarrow \cdots$$

Before taking homology, we must make our notation consistent. Relabel the modules to get

$$O \longrightarrow TP_o \xrightarrow{Td_1} TP_{-1} \xrightarrow{Td_2} TP_{-2} \longrightarrow \cdots,$$

and now relabel the maps

$$\cdots \longrightarrow TP_{-n} \xrightarrow{\Delta_{-n}} TP_{-n-1} \longrightarrow \cdots ,$$

so that $\Delta_{-n} = Td_{n+1}$.

DEFINITION. If T is contravariant, define R^nT on modules A by

$$R^nT(A) = H_{-n}(T\mathbf{P}_A) = \ker \Delta_{-n}/\mathrm{im}\, \Delta_{-n+1}$$

In terms of the original notation of the resolution,

$$R^nTA = \ker Td_{n+1}/\mathrm{im}\, Td_n .$$

Exercises. 6.15 Supply the definition of $R^nT(f)$, where $f\colon A \to B$ and T is contravariant. Prove that R^nT is a contravariant additive functor.

6.16. If T is contravariant and $\hat{R}^nT$ is defined as R^nT via different choices of projective resolutions, then R^nT and $\hat{R}^nT$ are naturally equivalent.

DEFINITION. If $T = \mathrm{Hom}_R(\ \ , A)$, then $R^nT = \mathrm{Ext}_R^n(\ \ , A)$. In particular,

$$\mathrm{Ext}_R^n(C, A) = \frac{\ker \mathrm{Hom}(d_{n+1}, A)}{\mathrm{im}\, \mathrm{Hom}(d_n, A)} ,$$

where $\cdots \to P_2 \xrightarrow{d_2} P_1 \xrightarrow{d_1} P_o \longrightarrow C \longrightarrow O$ is the chosen projective resolution of C.

COROLLARY 6.10. The definition of $\mathrm{Ext}_R^n(A, B)$ is independent of the choice of projective resolution of A.

Again, we ask whether the value of $\mathrm{Ext}_R^n(C,\ \ )$ on A is the same as the value of $\mathrm{Ext}_R^n(\ \ , A)$ on C. We answer this at the end of the chapter.

Résumé: $L_n T$ if T is covariant, use projective resolutions.
$R^n T$: if T is covariant, use injective-resolutions; if T is contravariant, use projective resolutions.

The reader may wonder why we have not considered left derived functors of Hom and right derived functors of $\otimes$. The reason is that we want to use the derived functors of T to investigate T, and this investigation is most fruitful when either $L_o T = T$ or $R^o T = T$. For any functor T, however, $L_o T$ is right exact and $R^o T$ is left exact, and so L_n Hom, for example, conveys no obvious information about Hom.

We want to consider how much exactness the derived functors preserve, and for this, we need a lemma that will enable us to use the exact triangle.

LEMMA 6.11. *Consider the diagram*

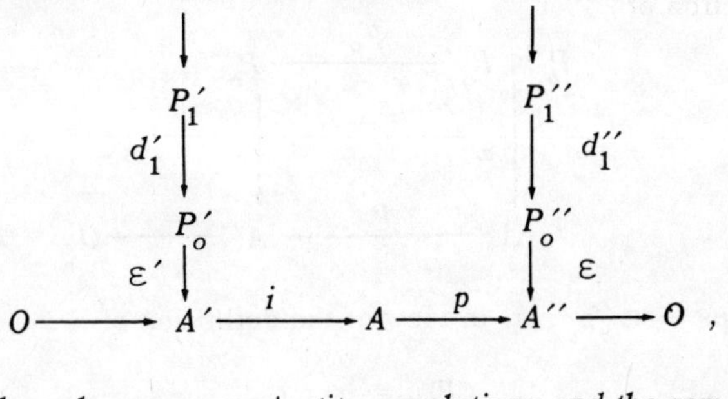

where the columns are projective resolutions and the row is exact. Then we may construct a projective resolution of A and chain maps so that the columns form an exact sequence of complexes.

The dual theorem holds as well.

Proof: Define $P_n = P'_n \oplus P''_n$; $i_n : P'_n \to P_n$ by $x' \mapsto (x', O)$, and $p_n : P_n \to P''_n$ by $(x', x'') \mapsto x''$. Clearly

$$O \longrightarrow P'_n \xrightarrow{i_n} P_n \xrightarrow{p_n} P''_n \longrightarrow O$$

is exact, for each n, and P_n is projective, being a sum of projectives. It remains to define maps $d_n : P_n \to P_{n-1}$ and $\varepsilon : P_o \to A$ so that all commutes and the center column is exact; these maps are essentially $d'_n + d''_n$.

Let us suppose we have a family of maps $\gamma_n : P''_n \to P'_{n-1}$, $n > O$, and $\gamma_o : P''_o \to A$, upon which further conditions will be imposed. Define

$$\varepsilon : P_o \longrightarrow A \text{ by } x'_o + x''_o \longmapsto i\varepsilon' x'_o + \gamma_o x''_o \ ;$$

$$d_n : P_n \longrightarrow P_{n-1} \text{ by } x'_n + x''_n \longmapsto d'_n x'_n + d''_n x''_n + \gamma_n x''_n \ .$$

With these definitions, one checks that no matter how the γ_n are chosen, every square in the new diagram commutes, with the possible exception of

$$\begin{array}{ccccc} P'_o \oplus P''_o & \xrightarrow{p_o} & P''_o & & \\ \downarrow \varepsilon & & \downarrow \varepsilon'' & & \\ A & \xrightarrow{p} & A'' & \longrightarrow & O \end{array}$$

We want $p\varepsilon = \varepsilon'' p_o$, or, in view of the definition of ε,

$$p\gamma_o = \varepsilon'' \ .$$

We also want the middle column to be a complex, i.e., $\varepsilon d_1 = O$ and $d_{n-1} d_n = O$. Substituting, these equations become

$$- i\varepsilon' \gamma_1 = \gamma_o d''_1$$

and

$$-d'_{n-1}\gamma_n = \gamma_{n-1}d''_n\,, \qquad n \geq 2\,.$$

Consider the following diagram:

$$\begin{array}{ccccccccccc}
\cdots \longrightarrow & P''_2 & \xrightarrow{d''_2} & P''_1 & \xrightarrow{d''_1} & P''_0 & \xrightarrow{\varepsilon''} & A'' & \longrightarrow & O \\
 & & & & & & & \downarrow 1_{A''} & & \\
\cdots \longrightarrow & P'_1 & \xrightarrow{-d'_1} & P'_0 & \xrightarrow{-i\varepsilon'} & A & \xrightarrow{p} & A'' & \longrightarrow & O
\end{array}$$

Since each term in the top row (except A'') is projective, and since the bottom row is exact, the comparison theorem provides maps $\gamma_n\colon P''_n \to P'_{n-1}$ making the resulting diagram commute, i.e., the equations we want can be satisfied. Finally, we must show that the middle column is exact. But we have just constructed an exact sequence of complexes.

$$O \longrightarrow \mathbf{P}' \longrightarrow \mathbf{P} \longrightarrow \mathbf{P}'' \longrightarrow O\,,$$

so that the exact triangle gives exactness of

$$\cdots \longrightarrow H_n(\mathbf{P}') \longrightarrow H_n(\mathbf{P}) \longrightarrow H_n(\mathbf{P}'') \longrightarrow \cdots$$

Since $\mathbf{P}'$ and $\mathbf{P}''$ are exact, $H_n(\mathbf{P}') = O = H_n(\mathbf{P}'')$; therefore $H_n(\mathbf{P}) = O$ for all n, i.e., the middle column is exact.

REMARK: It follows that $O \to \mathbf{P}_{A'} \to \mathbf{P}_A \to \mathbf{P}_{A''} \to O$ is also an exact sequence of complexes.

THEOREM 6.12. *Let $O \to A' \to A \to A'' \to O$ be an exact sequence of modules. If T is covariant, there is an exact sequence*

$$\cdots \longrightarrow L_nTA' \longrightarrow L_nTA \longrightarrow L_nTA'' \xrightarrow{\partial} L_{n-1}TA' \longrightarrow \cdots$$

Proof: Let $\cdots \to P_1' \to P_0' \to A' \to O$ and $\cdots \to P_1'' \to P_0'' \to A'' \to O$ be the chosen projective resolutions of A' and A''. We are now in the situation of Lemma 6.11, so there is a projective resolution of A, $\cdots \to \hat{P}_1 \to \hat{P}_0 \to A \to O$, that we may fit in the middle. Upon deleting, we have an exact sequence of complexes

$$O \longrightarrow \mathbf{P}_{A'} \longrightarrow \hat{\mathbf{P}}_A \longrightarrow \mathbf{P}_{A''} \longrightarrow O\,.$$

Applying T gives another exact sequence of complexes

$$O \longrightarrow T\mathbf{P}_{A'} \longrightarrow T\hat{\mathbf{P}}_A \longrightarrow T\mathbf{P}_{A''} \longrightarrow O\ ,$$

for each row $O \to P_n' \to \hat{P}_n \to P_n'' \to O$ is split and exercise 6.4 applies. (In general, the exact sequence of complexes is not split, for there is no reason why the splittings $P_n'' \to \hat{P}_n$ should constitute a chain map $\mathbf{P}_{A''} \to \hat{\mathbf{P}}_A$.) The exact triangle gives an exact sequence

$$\cdots \longrightarrow H_n(T\mathbf{P}_{A'}) \longrightarrow H_n(T\hat{\mathbf{P}}_A) \longrightarrow H_n(T\mathbf{P}_{A''}) \xrightarrow{\ \partial\ } H_{n-1}(T\mathbf{P}_{A'}) \longrightarrow \cdots$$

i.e., an exact sequence

$$\cdots \longrightarrow L_nTA' \longrightarrow \hat{L}_nTA \longrightarrow L_nTA'' \longrightarrow L_{n-1}TA' \longrightarrow \cdots$$

Notice that we have $\hat{L}_nTA$ instead of L_nTA, because the projective resolution we constructed with Lemma 6.11 need not be the projective resolution of A chosen originally. But this is no cause for concern; Theorem 6.7 gives $\hat{L}_nT$ and L_nT naturally equivalent. There is thus an exact sequence

$$\cdots \longrightarrow L_nTA' \longrightarrow L_nTA \longrightarrow L_nTA'' \longrightarrow L_{n-1}TA' \longrightarrow \cdots$$

With minor changes, the reader may prove analogous theorems for right derived functors.

THEOREM. 6.13. *If* $O \to A' \to A \to A'' \to O$ *is an exact sequence of modules and if* T *is covariant, there is an exact sequence*

$$\cdots \longrightarrow R^nTA' \longrightarrow R^nTA \longrightarrow R^nTA'' \xrightarrow{\partial} R^{n+1}TA' \longrightarrow \cdots$$

Proof: Only one point needs comment: why does $\partial: R^nTA'' \to R^{n+1}TA'$? If $\mathrm{E}_{A''}$ and $\mathrm{E}_{A'}$ are deleted injective resolutions, then $\partial: H_{-n}(T\mathrm{E}_{A''}) \to H_{-n-1}(T\mathrm{E}_{A'})$. Since $R^nTA' = H_{-n}(T\mathrm{E}_{A'})$, we have $H_{-n-1}(T\mathrm{E}_{A'}) = R^{n+1}TA'$.

THEOREM 6.14. *If* $O \to A' \to A \to A'' \to O$ *is an exact sequence of modules and if* T *is contravariant, there is an exact sequence*

$$\cdots \longrightarrow R^nTA'' \longrightarrow R^nTA \longrightarrow R^nTA' \longrightarrow R^{n+1}TA'' \longrightarrow \cdots$$

Our next aim is to prove the groups $\mathrm{Tor}_n^R(A, B)$ and $\mathrm{Ext}_R^n(A, B)$ are independent of the variable resolved. The idea is to compare each version (first variable resolved or second variable resolved) with a new version in which we resolve both variables simultaneously.

LEMMA 6.15. (Five lemma) *Consider the commutative diagram with exact rows*

$$\begin{array}{ccccccccc} A_1 & \longrightarrow & A_2 & \longrightarrow & A_3 & \longrightarrow & A_4 & \longrightarrow & A_5 \\ \downarrow{\scriptstyle \alpha_1} & & \downarrow{\scriptstyle \alpha_2} & & \downarrow{\scriptstyle \alpha_3} & & \downarrow{\scriptstyle \alpha_4} & & \downarrow{\scriptstyle \alpha_5} \\ B_1 & \longrightarrow & B_2 & \longrightarrow & B_3 & \longrightarrow & B_4 & \longrightarrow & B_5 \end{array}$$

(i) *If* α_2 *and* α_4 *are epic and* α_5 *is monic, then* α_3 *is epic;*

(ii) *If* α_2 *and* α_4 *are monic and* α_1 *is epic, then* α_3 *is monic.*

In particular, if α_1, α_2, α_4, *and* α_5 *are isomorphisms, so is* α_3.

Proof: Diagram-chasing. Note that (i) and (ii) are dual.

DEFINITION. If $\mathbf{A}$ and $\mathbf{A}'$ are complexes, their **tensor product** $\mathbf{A} \otimes \mathbf{A}'$ is the complex defined as follows:

$$(\mathbf{A} \otimes \mathbf{A}')_n = \sum_{i+j=n} A_i \otimes A'_j \ ;$$

$$\Delta_n\colon (\mathbf{A} \otimes \mathbf{A}')_n \longrightarrow (\mathbf{A} \otimes \mathbf{A}')_{n-1}$$

by

$$a_i \otimes a'_j \longmapsto da_i \otimes a'_j + (-1)^i a_i \otimes d'a'_j,$$

where $a_i \in A_i$, $a'_j \in A'_j$.

The sign $(-1)^i$ is needed in the definition of Δ_n to make $\Delta_{n-1}\Delta_n = O$.

Observe that we may consider a module A as a complex **concentrated in degree** O, i.e., $A_o = A$ and $A_n = O$ for $n \neq O$.

Note that $A \otimes \mathbf{A}' = \cdots \to A \otimes A'_n \xrightarrow{1 \otimes d'_n} A \otimes A'_{n-1} \to \cdots$; the definition above is thus consistent with that of $T\mathbf{A}'$, where $T = A \otimes$.

Exercise 6.17. Let $\cdots \to P_1 \to P_o \xrightarrow{\varepsilon} A \to O$ be an exact sequence.

(i) $H_n(\mathbf{P}_A) = O$ if $n > O$;

(ii) $H_o(\mathbf{P}_A) \cong A$;

(iii) $\varepsilon : \mathbf{P}_A \to A$ is a chain map (A concentrated in degree O);

(iv) For each n, $H_n(\varepsilon)\colon H_n(\mathbf{P}_A) \to H_n(A)$ is an isomorphism.

THEOREM 6.16. *If* $\cdots \to P_1 \to P_o \xrightarrow{\varepsilon} A \to O$ *and* $\cdots \to Q_1 \to Q_o \to B \to O$ *are projective resolutions, then*

$$H_n(A \otimes \mathbf{Q}_B) \cong H_n(\mathbf{P}_A \otimes \mathbf{Q}_B) \cong H_n(\mathbf{P}_A \otimes B) .$$

Thus, the two definitions of Tor_n^R *have the same value on* (A, B).

Proof: We define a "filtration" of the complex $A \otimes \mathbf{Q}_B$, i.e., an ascending sequence of subcomplexes

$$\mathbf{0} = \mathbf{G}^{-1} \subset \mathbf{G}^o \subset \mathbf{G}^1 \subset \cdots$$

as follows:

$$\text{in degree } n, \quad (\mathbf{G}^k)_n = \begin{cases} O & \text{if } n > k \\ A \otimes Q_n & \text{if } n \leq k \end{cases} .$$

Thus

$$\begin{aligned} \mathbf{G}^o &= \cdots \longrightarrow O \longrightarrow A \otimes Q_o \longrightarrow O ; \\ \mathbf{G}^1 &= \cdots \longrightarrow O \longrightarrow A \otimes Q_1 \longrightarrow A \otimes Q_o \longrightarrow O ; \\ \mathbf{G}^k &= \cdots \longrightarrow O \longrightarrow A \otimes Q_k \longrightarrow \cdots \longrightarrow A \otimes Q_o \longrightarrow O . \end{aligned}$$

Next, we define a filtration of $\mathbf{P}_A \otimes \mathbf{Q}_B$ as follows:

$$(\mathbf{F}^k)_n = \sum P_i \otimes Q_j, \quad \text{where } i + j = n \text{ and } j \leq k.$$

For each fixed k, $\varepsilon \otimes 1$ induces chain maps $\mathbf{F}^k \to \mathbf{G}^k$, namely, $\Sigma\, p_i \otimes q_j \mapsto \varepsilon p_o \otimes q_n$. There results a commutative diagram (of complexes) with exact rows

$$\begin{array}{ccccccccc} O & \longrightarrow & \mathbf{F}^{k-1} & \longrightarrow & \mathbf{F}^k & \longrightarrow & \mathbf{F}^k/\mathbf{F}^{k-1} & \longrightarrow & O \\ & & \downarrow & & \downarrow & & \downarrow & & \\ O & \longrightarrow & \mathbf{G}^{k-1} & \longrightarrow & \mathbf{G}^k & \longrightarrow & \mathbf{G}^k/\mathbf{G}^{k-1} & \longrightarrow & O \end{array} .$$

Using exercise 6.3, let us compute the quotient complexes.

$$\mathrm{G}^k/\mathrm{G}^{k-1} = A \otimes Q_k \text{ concentrated in degree } k\,;$$

$$\mathrm{F}^k/\mathrm{F}^{k-1} = \cdots \longrightarrow P_2 \otimes Q_k \longrightarrow P_1 \otimes Q_k \longrightarrow P_o \otimes Q_k \longrightarrow O\,,$$

where $P_o \otimes Q_k$ occurs in degree k.

Theorem 6.4 gives a commutative diagram with exact rows

$$\begin{array}{ccccccccc} H_{n+1}(\mathrm{F}^k/\mathrm{F}^{k-1}) & \to & H_n(\mathrm{F}^{k-1}) & \to & H_n(\mathrm{F}^k) & \to & H_n(\mathrm{F}^k/\mathrm{F}^{k-1}) & \to & H_{n-1}(\mathrm{F}^{k-1}) \\ \downarrow & & \downarrow & & \downarrow & & \downarrow & & \downarrow \\ H_{n+1}(\mathrm{G}^k/\mathrm{G}^{k-1}) & \to & H_n(\mathrm{G}^{k-1}) & \to & H_n(\mathrm{G}^k) & \to & H_n(\mathrm{G}^k/\mathrm{G}^{k-1}) & \to & H_{n-1}(\mathrm{G}^{k-1}) \end{array}$$

Since Q_k is projective, hence flat, we have exactness of

$$\cdots \longrightarrow P_1 \otimes Q_k \longrightarrow P_o \otimes Q_k \xrightarrow{\ \varepsilon \otimes 1\ } A \otimes Q_k \longrightarrow O\ .$$

It follows from exercise 6.17 that $\varepsilon \otimes 1$ induces isomorphisms $H_n(\mathrm{F}^k/\mathrm{F}^{k-1}) \to H_n(\mathrm{G}^k/\mathrm{G}^{k-1})$ for all n and k.

Let us prove, by double induction on n and k, that $\varepsilon \otimes 1$ induces isomorphisms $H_n(\mathrm{F}^k) \to H_n(\mathrm{G}^k)$. The induction begins, for both groups are O when n is negative and k is arbitrary; further, both groups are O for any n and $k = -1$. By induction, the second and fifth maps in the diagram above are isomorphisms, so the Five Lemma completes the induction.

Finally, if $k > n$, then $\mathrm{F}_n^k = (\mathrm{P}_A \otimes \mathrm{Q}_B)_n$ and $\mathrm{G}_n^k = (A \otimes \mathrm{Q}_B)_n$, so that $H_n(\mathrm{F}^k) = H_n(\mathrm{P}_A \otimes \mathrm{Q}_B)$ and $H_n(\mathrm{G}^k) = H_n(A \otimes \mathrm{Q}_B)$. An isomorphism $H_n(\mathrm{P}_A \otimes \mathrm{Q}_B) \cong H_n(\mathrm{P}_A \otimes B)$ is obtained similarly.

A dual proof works for Ext, i.e., with $\otimes$ replaced by Hom. It only remains to make $\mathrm{Hom}(\mathbf{A}, \mathbf{A}')$ into a complex, and we do this by first enlarging it so it contains more than chain maps.

DEFINITION. $\overline{\text{Hom}}_n(A, A') = \prod_{-\infty < i < \infty} \text{Hom}(A_i, A'_{i+n})$;

$$\Delta_n : \overline{\text{Hom}}_n(A, A') \longrightarrow \overline{\text{Hom}}_{n-1}(A, A')$$

by $$(f_i) \longmapsto (d'f_i + (-1)^n f_{i-1} d) \ .$$

Chain maps are thus certain maps of degree O, i.e., the old $\text{Hom}(A, A')$ is a subgroup of $\overline{\text{Hom}}_o(A, A')$. Homotopies are maps of degree $+1$; differentiations are maps of degree -1.

THEOREM 6.17. *If* $\cdots \to P_1 \to P_o \to A \to O$ *is a projective resolution and* $O \to B \to E_o \to E_1 \to \cdots$ *is an injective resolution, then*

$$H_{-n}(\text{Hom}(A, E_B)) \cong H_{-n}(\text{Hom}(P_A, E_B)) \cong H_{-n}(\text{Hom}(P_A, B)).$$

Thus, the two definitions of Ext^n_R *have the same value on* (A, B).

We end this chapter with a remark about *cohomology*, which usually means a contravariant homology theory. The derived functors of a contravariant functor are thus called cohomology. In this case, all the usual terminology is decorated with the prefix "co": cocycle, coboundary, cohomology. For example, the material in §5 is called *cohomology of groups*, for it arose by applying the contravariant functor $\text{Hom}_G(\ \ , A)$ to a complex. This clear distinction–homology covariant, cohomology contravariant–is often blurred, probably because of Theorem 6.17. The cohomology functor $\text{Ext}(\ \ , B)$ has the same value on A as the homology functor $\text{Ext}(A, \ \ )$ has on B. As a result, one often sees the derived functors of Hom (either variance) called cohomology.

§7. *Ext*

We want to examine Ext more closely. Recall the definition (on modules):

$$\mathrm{Ext}_R^n(A, B) = H_{-n}(\mathrm{Hom}_R(\mathbf{P}_A, B)) = H_{-n}(\mathrm{Hom}_R(A, \mathbf{E}_B))$$

where $\mathbf{P}_A$ is a deleted projective resolution of A and $\mathbf{E}_B$ is a deleted injective resolution of B.

THEOREM 7.1. *If n is negative,* $\mathrm{Ext}^n(A, B) = O$ for all A, B.

Proof: If $\mathbf{E}_B = \cdots \to O \xrightarrow{d_2} O \xrightarrow{d_1} E_o \xrightarrow{d_o} E_{-1} \xrightarrow{d_{-1}} E_{-2} \to \cdots$, then $\mathrm{Hom}(A, \mathbf{E}_B)$ has only O's to the left of $\mathrm{Hom}(A, E_o)$, and so all homology groups are O there.

THEOREM 7.2. $\mathrm{Ext}^o(A,\ \ )$ *is naturally equivalent to* $\mathrm{Hom}(A,\ \ )$.

Proof: If $\mathbf{E}_B = \cdots \to O \xrightarrow{d_1} E_o \xrightarrow{d_o} E_{-1} \to \cdots$, then

$$\mathrm{Ext}^o(A, B) = \ker d_{o*} / \mathrm{im}\, d_{1*} = \ker d_{o*}\ .$$

If the (non-deleted) resolution is

$$O \longrightarrow B \xrightarrow{\varepsilon} E_o \xrightarrow{d_o} E_{-1} \longrightarrow \cdots,$$

then left exactness of $\mathrm{Hom}(A,\ \ )$ gives an exact sequence

$$O \longrightarrow \mathrm{Hom}(A, B) \xrightarrow{\varepsilon_*} \mathrm{Hom}(A, E_o) \xrightarrow{d_{o*}} \mathrm{Hom}(A, E_{-1})\ ,$$

so that $\ker d_{o*} = \mathrm{im}\,\varepsilon_*$, i.e., $\varepsilon_*\colon \mathrm{Hom}(A, B) \to \mathrm{Ext}^o(A, B)$ is an isomorphism. We let the reader check that the maps ε_* (one for each B) define a natural equivalence.

THEOREM 7.3. *If* $O \to B' \to B \to B'' \to O$ is an exact sequence *of modules, then there is an (infinitely long) exact sequence*

$$O \longrightarrow \mathrm{Hom}(A, B') \longrightarrow \mathrm{Hom}(A, B) \longrightarrow \mathrm{Hom}(A, B'') \xrightarrow{\partial} \mathrm{Ext}^1(A, B') \longrightarrow \cdots$$

Proof: Theorem 6.13.

Observe that Ext thus repairs the exactness we may have lost when we applied Hom.

THEOREM 7.4. $\mathrm{Ext}^o(\ \ , B)$ *is naturally equivalent to* $\mathrm{Hom}(\ \ , B)$.

THEOREM 7.5. *If* $O \to A' \to A \to A'' \to O$ *is an exact sequence of modules, then there is an exact sequence*

$$O \longrightarrow \mathrm{Hom}(A'', B) \longrightarrow \mathrm{Hom}(A, B) \longrightarrow \mathrm{Hom}(A', B) \xrightarrow{\partial} \mathrm{Ext}^1(A'', B) \longrightarrow \cdots$$

Exercise 7.1. Suppose $A' \subset A$ and $f\colon A' \to B$. Call $\partial f \in \mathrm{Ext}^1(A/A', B)$ the **obstruction** of f. Prove that f can be extended to A if and only if its obstruction is O.

THEOREM 7.6. *If* P *is projective, then* $\mathrm{Ext}^n(P, B) = O$ for all B *and all* $n > O$.

Proof: We know the groups $\mathrm{Ext}^n(P, B)$ are independent of the choice of projective resolution of P. Since P is projective,

$$\cdots \longrightarrow O \longrightarrow O \longrightarrow P_o \xrightarrow{\varepsilon} P \longrightarrow O$$

is a projective resolution, where $P_o = P$ and $\varepsilon = 1_P$.

THEOREM 7.7. *If* E *is injective, then* $\mathrm{Ext}^n(A, E) = O$ *for all* A *and all* $n > O$.

DEFINITION: An **extension** of A by C is an exact sequence of R-modules

$$O \longrightarrow A \longrightarrow B \longrightarrow C \longrightarrow O .$$

THEOREM 7.8. *If* $\text{Ext}^1(C, A) = O$, *then every extension of A by C is split.*

Proof: Let $O \to A \xrightarrow{i} B \to C \to O$ be an extension. Applying $\text{Hom}(\ , A)$ gives exactness of

$$\text{Hom}(B, A) \xrightarrow{i^*} \text{Hom}(A, A) \longrightarrow \text{Ext}^1(C, A) = O \ ,$$

so that i^* is epic. In particular, there is a map $g \in \text{Hom}(B, A)$ with $1_A = i^*(g) = gi$, and this says the original sequence splits.

Later, we will prove the converse of Theorem 7.8. Now we give the converse of Theorem 7.6.

COROLLARY 7.9. *If* $\text{Ext}^1(C, A) = O$ *for all A, then C is projective; if* $\text{Ext}^1(C, A) = O$ *for all C, then A is injective.*

Proof: Theorems 3.11 and 3.18.

The next two results show that Ext behaves as Hom does on sums and products.

THEOREM 7.10. *For all n,* $\text{Ext}^n(\Sigma_k A_k, B) \cong \Pi_k \text{Ext}^n(A_k, B)$.

Proof: We do an induction on n, the case $n = O$ being Theorem 2.2. For each k, take an exact sequence

$$O \longrightarrow L_k \longrightarrow P_k \longrightarrow A_k \longrightarrow O \ ,$$

where P_k is projective. There results an exact sequence

$$O \longrightarrow \Sigma L_k \longrightarrow \Sigma P_k \longrightarrow \Sigma A_k \longrightarrow O \ ,$$

and ΣP_k is projective. This gives a commutative diagram with exact rows

$$\begin{array}{ccccccc}
\mathrm{Hom}(\Sigma P_k, B) & \rightarrow & \mathrm{Hom}(\Sigma L_k, B) & \rightarrow & \mathrm{Ext}^1(\Sigma A_k, B) & \rightarrow & \mathrm{Ext}^1(\Sigma P_k, B) = O \\
\downarrow & & \downarrow & & & & \\
\Pi\ \mathrm{Hom}(P_k, B) & \rightarrow & \Pi\ \mathrm{Hom}(L_k, B) & \rightarrow & \Pi\ \mathrm{Ext}^1(A_k, B) & \rightarrow & \Pi\ \mathrm{Ext}^1(P_k, B) = O
\end{array}$$

(the maps in the bottom row are just the maps of Theorem 7.5 at each coordinate). Diagram chasing provides a map

$$\mathrm{Ext}^1(\Sigma A_k, B) \rightarrow \Pi\ \mathrm{Ext}^1(A_k, B)$$

which must be an isomorphism. The theorem is thus true for $n = 1$ (many inductions involving Ext^n start slowly). If $n > 1$, we have the diagram

$$\begin{array}{ccccccc}
O = \mathrm{Ext}^{n-1}(\Sigma P_k, B) & \rightarrow & \mathrm{Ext}^{n-1}(\Sigma L_k, B) & \xrightarrow{\partial} & \mathrm{Ext}^n(\Sigma A_k, B) & \rightarrow & \mathrm{Ext}^n(\Sigma P_k, B) = O \\
& & \downarrow \alpha & & & & \\
O = \Pi\ \mathrm{Ext}^{n-1}(P_k, B) & \rightarrow & \Pi\ \mathrm{Ext}^{n-1}(L_k, B) & \xrightarrow{\partial'} & \Pi\ \mathrm{Ext}^n(A_k, B) & \rightarrow & \Pi\ \mathrm{Ext}^n(P_k, B) = O
\end{array}$$

By exactness, ∂ and ∂' are isomorphisms (we use Theorem 7.6 again), while α is an isomorphism that exists by induction. It follows that $\partial' \alpha \partial^{-1}$: $\mathrm{Ext}^n(\Sigma A_k, B) \rightarrow \Pi\ \mathrm{Ext}^n(A_k, B)$ is an isomorphism.

THEOREM 7.11. *For all n,* $\mathrm{Ext}^n(A, \Pi B_k) \cong \Pi\ \mathrm{Ext}^n(A, B_k)$.

Proof: Dual to the proof of 7.10. In particular, begin with exact sequences $O \rightarrow B_k \rightarrow E_k \rightarrow Q_k \rightarrow O$, where E_k is injective, and then then apply Ext to $O \rightarrow \Pi B_k \rightarrow \Pi E_k \rightarrow \Pi Q_k \rightarrow O$.

Given a complex of R-modules and R-maps, the complex we obtain after applying Hom_R is a complex of abelian groups and Z-maps. If R is commutative, however, we may consider modules as

2-sided R-modules, and equip each Hom_R with an R-module structure. Furthermore, the maps between the Hom's are then R-maps. Conclusion:

THEOREM 7.12. *If R is commutative, $\mathrm{Ext}_R^n(A, B)$ is an R-module.*

Proof: Ext^n is just a quotient of two R-modules.

Let R be commutative and let A be an R-module. If $r \in R$, then $\mu: A \to A$ defined by $a \mapsto ra$ is an R-map, called **multiplication** (or *homothety*) by r. (Note that μ is not an R-map if r is not in the center of R.)

THEOREM 7.13. *If $\mu: A \to A$ is multiplication by r, where r is in the center of R, then $\mu^*: \mathrm{Ext}_R^n(A, B) \to \mathrm{Ext}_R^n(A, B)$ is also multiplication by r. This is also true for the second variable.*

Proof: There is a diagram

$$\begin{array}{ccccccccc} \cdots & \longrightarrow & P_1 & \longrightarrow & P_o & \longrightarrow & A & \longrightarrow & O \\ & & & & & & \big\downarrow{\scriptstyle \mu} & & \\ \cdots & \longrightarrow & P_1 & \longrightarrow & P_o & \longrightarrow & A & \longrightarrow & O \end{array}$$

where each row is (the same) projective resolution of A. Recall the definition of $\mu^*: \mathrm{Ext}^n(A, B) \to \mathrm{Ext}^n(A, B)$; first fill in a chain map g over μ (so that $g_n: P_n \to P_n$), then apply the functor $\mathrm{Hom}_R(\ \ , B)$ to the diagram, and

$$\mu^*(z_n + \text{boundaries}) = g_n z_n + \text{boundaries}.$$

We also know that any choice of chain map g over μ gives the same μ^*. In particular, if we define g by letting $g_n: P_n \to P_n$ be multiplication by r, then g is a chain map over μ, and

$$\mu^*(z_n + \text{boundaries}) = rz_n + \text{boundaries} = r(z_n + \text{boundaries}).$$

Exercises 7.2. Let R be a domain and A an R-module. We say that a multiplication by r is *nontrivial* if $r \neq O$. Prove that A is torsion-free if and only if every nontrivial multiplication is monic; prove that A is divisible if and only if every nontrivial multiplication is epic.

7.3. Let R be a domain, and let A be a divisible R-module. If $rA = O$ for some $r \neq O$, then $A = O$.

7.4. Let R be a domain; a functor $T: {}_R\mathfrak{M} \to {}_R\mathfrak{M}$ **preserves multiplications** if, whenever $\mu: A \to A$ is multiplication by r, then $T\mu: TA \to TA$ is also multiplication by r. Prove that if A is torsion-free divisible, then TA is also torsion-free divisible.

7.5. Let R be commutative, A an R-module with $rA = O$ for some A. If T preserves multiplications, then $rTA = O$.

We seek an interpretation of $\text{Ext}^1(C, A)$, and we begin with a definition motivated by Theorem 5.9.

DEFINITION. Two extensions ξ and ξ' of A by C are **equivalent** if there is a commutative diagram

$$\begin{array}{llllllllll} \xi: & O & \longrightarrow & A & \longrightarrow & E & \longrightarrow & C & \longrightarrow & O \\ & & & \downarrow 1_A & & \downarrow \phi & & \downarrow 1_C & & \\ \xi': & O & \longrightarrow & A & \longrightarrow & E' & \longrightarrow & C & \longrightarrow & O \end{array}.$$

It is easy to check that equivalence is, in fact, an equivalence relation (to prove symmetry, use the Five Lemma to show that ϕ is an isomorphism). The equivalence class of an extension $\xi: O \to A \to B \to C \to O$ is denoted $[\xi]$, and the set of all equivalence classes is denoted $e(C, A)$. Our aim is to identify $e(C, A)$

with $\operatorname{Ext}^1(C, A)$ (thus providing the etymology of Ext), but we need some preparation.

Exercise 7.6. All split extensions of A by C are equivalent.

Suppose $\xi\colon O \to A \to B \to C \to O$ is an extension. Given a projective resolution of C, we may form the diagram

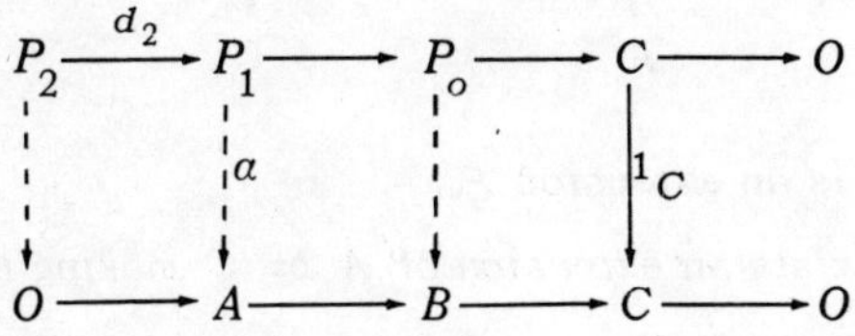

By the comparison theorem (6.6), we may fill in the dotted arrows, obtaining a commutative diagram. In particular, we have a map $\alpha\colon P_1 \to A$. Note that commutativity gives $\alpha d_2 = O$, so that $\alpha \in \ker d_2^*$ and α is a cocycle. Second, we know that α is unique to homotopy: if $\alpha'\colon P_1 \to A$ is a second such map, there are maps s_o, s_1 with $\alpha - \alpha' = O \cdot s_1 + s_o d_1$ (for there is only the zero map $O \to A$)

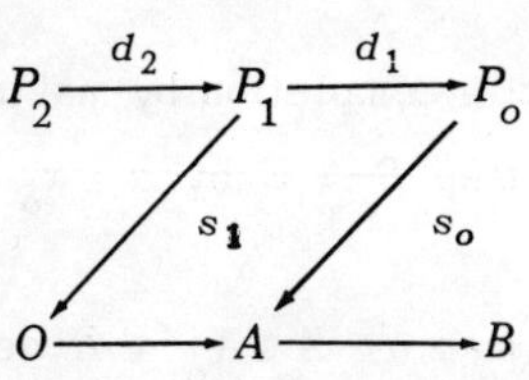

Thus $\alpha - \alpha' \in \operatorname{im} d_1^*$, and ξ determines a unique element of $\operatorname{Ext}^1(C, A)$. It is easy to check that if ξ and ξ' are equivalent, they determine the same element. In sum, we have defined a function $\psi\colon\ e(C, A) \to \operatorname{Ext}^1(C, A)$.

Exercise 7.7. If ξ is a split extension, then $\psi[\xi] = O$.

An analysis of the diagram occurring in the definition of ψ yields the following:

Lemma 7.14. *Consider the diagram*

$$\begin{array}{ccccccccc} 0 & \longrightarrow & X_1 & \xrightarrow{j} & X_o & \xrightarrow{\varepsilon} & C & \longrightarrow & 0 \\ & & \Big\downarrow{\scriptstyle \alpha} & & & & \Big\downarrow{\scriptstyle 1_C} & & \\ & & A & & & & C & & \end{array}$$

where the row is an extension ξ.

(i) *there exists an extension of* A *by* C *making the following diagram commute:*

$$\begin{array}{ccccccccc} 0 & \longrightarrow & X_1 & \xrightarrow{j} & X_o & \xrightarrow{\varepsilon} & C & \longrightarrow & 0 \\ & & \Big\downarrow{\scriptstyle \alpha} & & \Big\downarrow{\scriptstyle \beta} & & \Big\downarrow{\scriptstyle 1_C} & & \\ 0 & \longrightarrow & A & \xrightarrow{i} & B & \longrightarrow & C & \longrightarrow & 0 \end{array}$$

(ii) *In any completed diagram, the first square is a pushout.*

(iii) *Any two extensions completing the diagram are equivalent.*

Proof: (i) Construct a completion by making the first square a pushout; then define a map $B \to C$ by $a + x_o + S \mapsto \varepsilon x_o$, where $B = (A \oplus X_o)/S$.

(ii) First show the map $\sigma: A \oplus X_o \to B$ defined by $a + x_o \mapsto ia + \beta x_o$ is epic; then show $\ker \sigma = \{(\alpha x_1, -jx_1): x_1 \in X_1\}$.

(iii) Any two extensions completing the diagram begin with A and end with C. Since we have pushouts, the map between the middle modules is provided by Theorem 3.16.

Notation: The extension just constructed is denoted $\alpha\xi$.

Observe that the dual of Lemma 7.14 is also true. Now we begin with

$$\begin{array}{ccccccccc} & & A & & & & C & & \\ & & \downarrow{\scriptstyle 1_A} & & & & \downarrow{\scriptstyle \gamma} & & \\ \xi: O & \longrightarrow & A & \longrightarrow & Y_o & \longrightarrow & Y_1 & \longrightarrow & O\ . \end{array}$$

There does exist a completion, unique to equivalence, and the second square is a pullback.

Notation: The extension just constructed is denoted by $\xi\gamma$.

THEOREM 7.15. *The function* ψ: $e(C, A) \to \mathrm{Ext}^1(C, A)$ *is a one-to-one correspondence.*

Proof: We exhibit a function θ: $\mathrm{Ext}^1(C, A) \to e(C, A)$ inverse to ψ. If $\alpha : P_1 \to A$ is a cocycle, then $\alpha d_2 = O$ and α induces a map $\bar{\alpha} : P_1/\mathrm{im}\ d_2 \to A$. If Ξ is the extension

$$\Xi: O \longrightarrow P_1/\mathrm{im}\ d_2 \longrightarrow P_o \longrightarrow C \longrightarrow O\ ,$$

define θ by $\alpha + \mathrm{im}\ d_1^* \to [\bar{\alpha}\ \Xi]$. Thus we have the commutative diagram

$$\begin{array}{ccccccccc} O & \longrightarrow & P_1/\mathrm{im}\ d_2 & \xrightarrow{\bar{d}_1} & P_o & \longrightarrow & C & \longrightarrow & O \\ & & \downarrow{\scriptstyle \bar{\alpha}} & & \downarrow{\scriptstyle \beta} & & \downarrow{\scriptstyle 1_C} & & \\ O & \longrightarrow & A & \xrightarrow{i} & B & \longrightarrow & C & \longrightarrow & O \end{array}$$

Let us show θ is independent of the choice of representative α. Suppose $\alpha' = \alpha + sd_1$, where $s : P_o \to A$. The following diagram commutes:

$$\begin{array}{ccccccccc}
P_2 & \xrightarrow{d_2} & P_1 & \xrightarrow{d_1} & P_o & \longrightarrow & C & \longrightarrow & O \\
\downarrow{\scriptstyle \alpha+sd_1} & & \downarrow & & \downarrow{\scriptstyle \beta+is} & & \downarrow{\scriptstyle 1_C} & & \\
O & \longrightarrow & A & \xrightarrow{i} & B & \longrightarrow & C & \longrightarrow & O
\end{array}$$

and this shows that θ assigns the same value to α and $\alpha + sd_1$.

We claim $\psi\theta = 1$. Starting with $\alpha + \text{im } d_1^*$, we get the diagram

$$\begin{array}{ccccccccc}
O & \longrightarrow & P_1/\text{im } d_2 & \longrightarrow & P_o & \longrightarrow & C & \longrightarrow & O \\
 & & \downarrow{\scriptstyle \bar{\alpha}} & & \downarrow & & \downarrow{\scriptstyle 1_C} & & \\
O & \longrightarrow & A & \longrightarrow & B & \longrightarrow & C & \longrightarrow & O\ .
\end{array}$$

Visibly, ψ of the bottom row is $\alpha + \text{im } d_1^*$.

Finally, we claim $\theta\psi = 1$. Starting with $[\xi]$, we have $\psi[\xi] = \alpha + \text{im } d_1^*$, as in the diagram just above. Now θ sends $\alpha + \text{im } d_1^*$ into $[\bar{\alpha}\Xi]$, and, by Lemma 7.14, $[\bar{\alpha}\Xi] = [\xi]$.

COROLLARY 7.16. $\text{Ext}^1(C, A) = O$ *if and only if every extension of* A *by* C *is split.*

Proof: One way is Theorem 7.8. Conversely, if every extension is split, then exercise 7.6 implies $e(C, A)$ is a set with one element, so the theorem gives $\text{Ext}^1(C, A) = O$.

We have been treating $\text{Ext}^1(C, A)$ via projective resolutions of C; we could also treat it via injective resolutions of A. This approach yields a function ψ': $e(C, A) \to \text{Ext}^1(C, A)$ defined by $[\xi] \longmapsto \gamma + \text{im } d_{o*}$, where

$$\begin{array}{cccccccccc}
\xi: & O & \longrightarrow & A & \longrightarrow & B & \longrightarrow & C & \longrightarrow & O \\
 & & & \downarrow{\scriptstyle 1_A} & & \downarrow & & \downarrow{\scriptstyle \gamma} & & \downarrow \\
 & O & \longrightarrow & A & \longrightarrow & E_o & \xrightarrow{d_o} & E_1 & \longrightarrow & E_2 \longrightarrow \cdots .
\end{array}$$

Just as above, ψ' can be shown to be a one-to-one correspondence; its inverse is θ', where $\theta'(\gamma + \text{im } d_{o*}) = [\Omega\bar{\gamma}]$,

$$\Omega = O \longrightarrow A \longrightarrow E_o \longrightarrow E_1/\text{im } d_o \longrightarrow O$$

and

$$\bar{\gamma}: C \longrightarrow E_1/\text{im } d_o$$

is the map induced by the cocycle γ.

THEOREM 7.17. *Let A be an abelian group and let Z_n be cyclic of order n. Then* $\text{Ext}^1(Z_n, A) \cong A/nA$.

Proof: Apply Hom(, A) to $O \to Z \xrightarrow{n} Z \to Z_n \to O$, where the first map is multiplication by n, to obtain exactness of

$$\text{Hom}(Z, A) \xrightarrow{n} \text{Hom}(Z, A) \longrightarrow \text{Ext}^1(Z_n, A) \longrightarrow \text{Ext}^1(Z, A) \ .$$

Since Z is projective, the last Ext = O; also, the first map is still multiplication by n, while $\text{Hom}(Z, A) = A$. It follows that $\text{Ext}^1(Z_n, A) \cong A/nA$.

We can now tie a loose end from §5. There $e(C, A)$ meant all extensions (not necessarily abelian) of the abelian group A by the group C. Consider the special case $C \cong Z_p \cong A$, where p is prime. Since the middle group of any extension has order p^2, it is necessarily abelian. Thus $e(Z_p, Z_p)$ is unambiguous and is $\text{Ext}_Z^1(Z_p, Z_p)$. By Theorem 7.17, $e(Z_p, Z_p) \cong Z_p$.

Exercises 7.8. If A is an abelian group with $nA = A$, then every exact sequence $O \to A \to E \to Z_n \to O$ splits.

7.9. If A is a torsion abelian group, prove that $\text{Ext}_Z^1(A, Z) \cong \text{Hom}(A, T)$, where T is the circle group, i.e., $T = R/Z$, where R is the additive group of reals.

7.10 Let R be any ring, A a left R-module, and suppose $nA = O$ for some integer n. If $T: {}_R\mathfrak{M} \to {}_S\mathfrak{M}$ is an additive functor, then $n(TA) = O$.

If G is a group, X a set, and $\psi: X \to G$ is a one-to-one correspondence, there is a unique addition on X making X a group and ψ an isomorphism. What is the addition in $e(C, A)$? First we do some bookkeeping; in fact, we have three bookkeeping formulas.

(1) If $[\xi] \in e(C, A)$ and $h: A \to A'$, then

$$\psi[h\xi] = h\psi[\xi].$$

This follows from the diagram

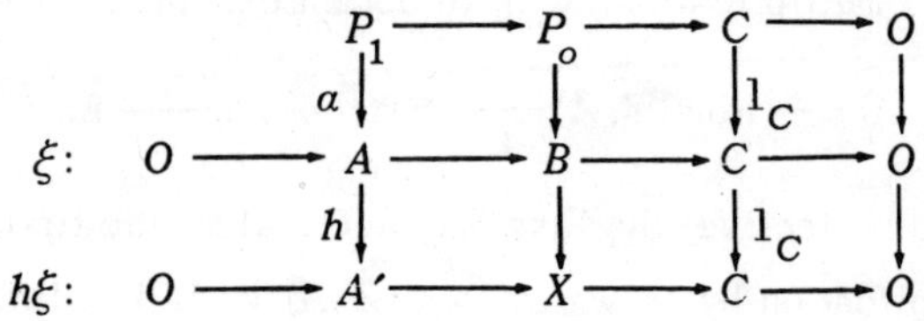

(2) If $[\xi] \in e(C, A)$ and $k: C' \to C$, then

$$\psi[\xi k] = \psi[\xi]\bar{k}_1,$$

where $\bar{k}_1$ is part of a chain map over k (see below for details). Consider the diagram

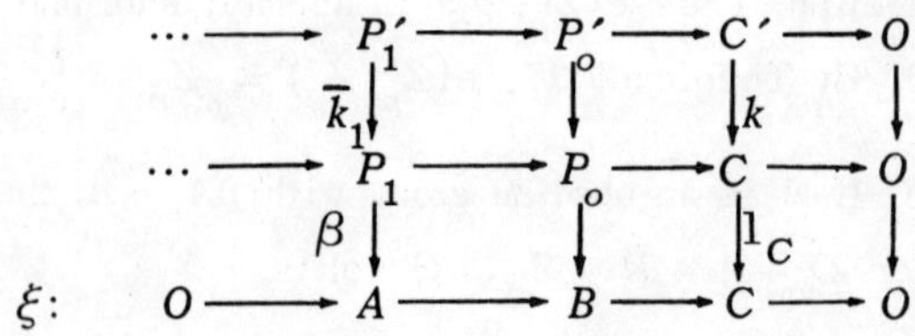

Now $\bar{k}_1$ is as advertised and $\psi[\xi]$ is the coset of β. To compute $\psi[\xi k]$, consider the following diagram.

$$\begin{array}{lccccccccc}
 & \cdots & \longrightarrow & P_1' & \longrightarrow & P_o' & \longrightarrow & C' & \longrightarrow & 0 \\
 & & & \downarrow{\scriptstyle a} & & \downarrow & & \downarrow{\scriptstyle 1_{C'}} & & \downarrow \\
\xi k: & 0 & \longrightarrow & A & \longrightarrow & X & \longrightarrow & C' & \longrightarrow & 0 \\
 & & & \downarrow{\scriptstyle 1_A} & & \downarrow & & \downarrow{\scriptstyle k} & & \downarrow \\
\xi: & 0 & \longrightarrow & A & \longrightarrow & B & \longrightarrow & C & \longrightarrow & 0
\end{array}$$

Thus, $\psi[\xi k]$ is the coset of a. Since both $\beta \bar{k}_1$ and a are the first part of a chain map over k, uniqueness to homotopy puts them in the same coset.

The most obvious way to put two extensions ξ and ξ' together is to form their **direct sum** $\xi \oplus \xi'$:

$$0 \longrightarrow A \oplus A' \xrightarrow{i \oplus i'} B \oplus B' \xrightarrow{\pi \oplus \pi'} C \oplus C' \longrightarrow 0$$

(3) $\psi[\xi \oplus \xi'] = \psi[\xi] \oplus \psi[\xi']$.

This follows from the diagram

$$\begin{array}{ccccccccc}
\cdots & \longrightarrow & P_1 \oplus P_1' & \longrightarrow & P_o \oplus P_o' & \longrightarrow & C \oplus C' & \longrightarrow & 0 \\
 & & \downarrow & & \downarrow & & \downarrow{\scriptstyle 1} & & \\
0 & \longrightarrow & A \oplus A' & \longrightarrow & B \oplus B' & \longrightarrow & C \oplus C' & \longrightarrow & 0 ,
\end{array}$$

for the top row is a projective resolution of $C \oplus C'$.

We are now in a position to construct the sum of two extensions in $e(C, A)$. First, form the direct sum

$$0 \longrightarrow A \oplus A \longrightarrow B \oplus B' \longrightarrow C \oplus C \longrightarrow 0 .$$

To get an extension beginning with A, we want a map $\nabla : A \oplus A \to A$, which gives us an extension in $e(C \oplus C, A)$. Finally, we want a map $\Delta : C \to C \oplus C$, which will give an extension in $e(C, A)$. The clue to good choices of ∇ and Δ is the remark that Ext is constructed from Hom. Define $\Delta : X \to X \oplus X$ by $\Delta x = (x, x)$ and define

$\nabla : Y \oplus Y \to Y$ by $\nabla(y, y') = y + y'$. If $f, g : X \to Y$ are homomorphisms, then $\nabla(f \oplus g)\Delta : X \to Y$ and sends

$$x \longmapsto (x, x) \longmapsto (fx, gx) \longmapsto fx + gx .$$

Therefore $\nabla(f \oplus g)\Delta = f + g$.

Exercise 7.11. Consider the diagram

$$\begin{array}{ccccccccc} \cdots \longrightarrow & P_1 & \xrightarrow{d_1} & P_o & \xrightarrow{\varepsilon} & C & \longrightarrow & O \\ & & & & & \downarrow \Delta & & \\ \longrightarrow & P_1 \oplus P_1 & \xrightarrow{d_1 \oplus d_1} & P_o \oplus P_o & \xrightarrow{\varepsilon \oplus \varepsilon} & C \oplus C & \longrightarrow & O \end{array}$$

Prove the diagram commutes if we fill in $\Delta : P_n \to P_n \oplus P_n$ at each n (a similar result holds for ∇). Conclude that we may take $\bar{\Delta}_1 = \Delta$ in the bookkeeping formula.

THEOREM 7.18. *e(C, A) is an abelian group under* **Baer sum,**

$$[\xi] + [\xi'] = [\nabla(\xi \oplus \xi')\Delta],$$

and $\psi : e(C, A) \to \operatorname{Ext}^1(C, A)$ *is an isomorphism.*

REMARK. It is true that associativity $[(\alpha\xi)\gamma] = [\alpha(\xi\gamma)]$ holds; however, we do not need nor use this fact in this proof.

Proof: The formula for Baer sum defines a relation

$$\rho : e(C, A) \times e(C, A) \longrightarrow e(C, A) ,$$

which we do not yet know is a function. The bookkeeping formulas give

$$\begin{aligned} \psi[\nabla(\xi \oplus \xi')\Delta] &= \psi[\nabla(\xi \oplus \xi')]\bar{\Delta}_1 \\ &= \nabla\psi[\xi \oplus \xi']\bar{\Delta}_1 \end{aligned}$$

$$= \nabla(\psi[\xi] \oplus \psi[\xi'])\bar{\Delta}_1$$
$$= \nabla(\psi[\xi] \oplus \psi[\xi'])\Delta, \quad \text{by exercise 7.11}$$
$$= \psi[\xi] + \psi[\xi'] \ .$$

There are two conclusions from this computation. First, $\psi\rho$ is a function, so that, since ψ is a one-to-one correspondence, $\rho = \psi^{-1}(\psi\rho)$ is also a function. Second, $\psi([\xi] + [\xi']) = \psi[\xi] + \psi[\xi']$, so ρ is the good addition on $e(C, A)$.

COROLLARY 7.19. *Let* Ξ *be the extension*

$$O \longrightarrow P_1/\text{im } d_2 \longrightarrow P_o \longrightarrow C \longrightarrow O \ .$$

Then $[(a + a')\Xi] = [a\Xi] + [a'\Xi]$, *where* $a, a'\colon P_1/\text{im } d_2 \to A$.

Proof: Take ψ of both sides.

This Corollary gives a second look at addition in $e(C, A)$. In a certain sense, it is "cyclic with generator Ξ" for the fact that $\psi^{-1} = \theta$ is onto says that every element of $e(C, A)$ is of the form $[a\Xi]$. Everything dualizes when we take injective resolutions.

COROLLARY 7.20. *Let* Ω *be the extension*

$$O \longrightarrow A \longrightarrow E_o \longrightarrow E_1/\text{im } d_o \longrightarrow O \ .$$

Then $[\Omega(\gamma + \gamma')] = [\Omega\gamma] + [\Omega\gamma']$, *where* $\gamma, \gamma'\colon C \to E_1/\text{im } d_o$.

Proof: Take ψ' of both sides, using the analogue of Theorem 7.18.

COROLLARY 7.21. *If* $[\xi] \in e(C, A)$, *then there are maps* a *and* γ *with* $[a\Xi] = [\xi] = [\Omega\gamma]$. *The function* $\tau\colon \text{Ext}^1(C, A) \to \text{Ext}^1(C, A)$ *defined by*

$$\alpha + \operatorname{im} d_1^* \longmapsto \gamma + \operatorname{im} d_{o*}$$

is an isomorphism.

REMARK. This is the poor man's version of Theorem 6.17.

Proof: The reader may check τ is a one-to-one correspondence; τ is an isomorphism by Corollaries 7.19 and 7.20.

We leave further properties of $e(C, A)$ to the exercises. The reader will see there an elegant description of induced maps and the connecting homomorphism.

Exercises 7.12. Show that $-[\xi] = [(-1_A)\xi] = [\xi(-1_C)]$.

7.13. $e(C,\ \ )$ is a functor: if $h: A \to A'$, define

$$h_\#: e(C, A) \longrightarrow e(C, A') \text{ by } [\xi] \longmapsto [h\xi].$$

Show that $e(C,\ \ )$ and $\operatorname{Ext}^1(C,\ \ )$ are naturally equivalent.

7.14. If χ is the extension $O \to A' \xrightarrow{i} A \xrightarrow{\pi} A'' \to O$, define a map $D: \operatorname{Hom}(C, A'') \to e(C, A')$ by $g \to [\chi g]$. Prove exactness of the sequence

$$\operatorname{Hom}(C, A) \xrightarrow{\pi_*} \operatorname{Hom}(C, A'') \xrightarrow{D} e(C, A') \xrightarrow{i_\#} e(C, A) \xrightarrow{\pi_\#} e(C, A$$

7.15. Use the previous two exercises to prove commutativity of

$$\begin{array}{ccc} & \xrightarrow{D} & e(C, A') \\ \operatorname{Hom}(C, A'') & & \downarrow \psi \\ & \xrightarrow{\partial} & \operatorname{Ext}^1(C, A'). \end{array}$$

7.16. $e(\ \ , A)$ is a contravariant functor: if $k: C' \to C$, then $k^\#: e(C, A) \to e(C', A)$ is given by $[\xi] \mapsto [\xi k]$. Prove that $e(\ \ , A)$ and $\operatorname{Ext}^1(\ \ , A)$ are naturally equivalent.

7.17. If Γ is the extension $0 \to C' \to C \to C'' \to 0$, define $D\colon \operatorname{Hom}(C', A) \to e(C'', A)$ by $g \mapsto [g\Gamma]$. Prove exactness of

$$\operatorname{Hom}(C, A) \longrightarrow \operatorname{Hom}(C', A) \xrightarrow{D} e(C'', A) \longrightarrow e(C, A) \longrightarrow e(C', A).$$

7.18. Using the previous two exercises, prove commutativity of

$$\begin{array}{ccc} & \overset{D}{\nearrow} & e(C'', A) \\ \operatorname{Hom}(C', A) & & \downarrow \psi \\ & \underset{\partial}{\searrow} & \operatorname{Ext}^1(C'', A)\,. \end{array}$$

Yoneda has shown how to generalize the construction of $e(C, A)$, thus giving interpretations of the higher Ext^n, induced maps, and connecting homomorphisms (see MacLane, p. 82).

A final remark. The naive approach to extensions that we followed in §5 can also be followed here, and, as there, it leads to a special free resolution of the first variable. Thus, given an extension of left R-modules $0 \to A \to B \to C \to 0$, let us choose a lifting $\lambda\colon C \to B$ with $\lambda 0 = 0$. Each element of B has a unique expression of the form $a + \lambda c$, and, when we try to add two elements, a factor set emerges:

$$(a + \lambda c) + (a' + \lambda c') = a + a' + f(c, c') + \lambda(c + c'),$$

where $f\colon C \times C \to A$. Such a function satisfies identities:

(i) $f(c, 0) = 0 = f(0, c)$;

(ii) $f(c', c'') - f(c + c', c'') + f(c, c' + c'') - f(c, c') = 0$;

(iii) $f(c, c') = f(c', c)$.

The second identity arises from associativity, the third arises from commutativity. This isn't enough, for such a function f only carries the information that B is an abelian group. To ensure that B is a module, observe that there is a second function $g\colon R \times C \to A$ defined by $g(r, c) = r\lambda c - \lambda rc$. There are more identities which come from the module axioms:

(iv) $g(1, c) = O = g(r, O)$;

(v) $rg(s, c) = g(rs, c) - g(r, sc)$;

(iv) $g(r+s, c) + f(rc, sc) = g(r, c) + g(s, c)$;

(vii) $g(r, c+c') + f(rc, rc') = g(r, c) + g(r, c')$.

The ordered pair (f, g) conveys all the data. We only remark that the set of all such (f, g) forms an abelian group (where each coordinate acts via pointwise addition), and that if one chooses a second lifting λ': $C \to B$, then he determines a subgroup. Obviously, the resolution and boundary formula are more complicated than before (although they are simple for Z-modules when g can be dispensed with), but a stubborn man with courage can write them down.

§8. *Tor*

Now we turn to Tor. First, recall the definition (on modules):

$$\mathrm{Tor}_n^R(A, B) = H_n(\mathbf{P}_A \otimes_R B) = H_n(A \otimes_R \mathbf{P}_B),$$

where $\mathbf{P}_A$ and $\mathbf{P}_B$ are deleted projective resolutions of A and B respectively.

THEOREM 8.1. $\mathrm{Tor}_n(A, B) = O$ *when* n *is negative, all* A, B.

Proof: If $\mathbf{P}_B = \cdots \to P_1 \to P_o \to O$, then $A \otimes \mathbf{P}_B$ has only O's to the right of $A \otimes P_o$.

THEOREM 8.2. $\mathrm{Tor}_o(A, \;)$ *is naturally equivalent to* $A \otimes \;$; *similarly in the other variable.*

Proof: If $\mathbf{P}_B = \cdots \to P_1 \xrightarrow{d_1} P_o \xrightarrow{d_o} O$, then

$$\mathrm{Tor}_o(A, B) = \ker 1 \otimes d_o / \mathrm{im}\; 1 \otimes d_1 = \mathrm{coker}\; 1 \otimes d_1 .$$

But if the (non-deleted) resolution is $\cdots \to P_1 \xrightarrow{d_1} P_o \xrightarrow{\varepsilon} B \to O$, then right exactness of $A \otimes$ gives an exact sequence

$$A \otimes P_1 \xrightarrow{1 \otimes d_1} A \otimes P_o \xrightarrow{1 \otimes \varepsilon} A \otimes B \longrightarrow O ,$$

so that $1 \otimes \varepsilon$ induces an isomorphism $\mathrm{coker}\; 1 \otimes d_1 \to A \otimes B$. We let the reader check that these maps define a natural equivalence.

THEOREM 8.3 *If* $O \to B' \to B \to B'' \to O$ *is an exact sequence of modules, there is an (infinitely long) exact sequence*

$$\cdots \longrightarrow \mathrm{Tor}_1(A, B'') \longrightarrow A \otimes B' \longrightarrow A \otimes B \longrightarrow A \otimes B'' \longrightarrow O;$$

similarly in the other variable.

Proof: Theorem 6.12.

Observe that Tor thus repairs the exactness we may have lost when we applied $\otimes$;

THEOREM 8.4. *If P is projective, then* $\operatorname{Tor}_n(P, B) = 0$ *for all B and for all $n > 0$; similarly in the other variable.*

Proof: We know the groups $\operatorname{Tor}_n(P, B)$ are independent of the choice of projective resolution of P. Since P is projective,

$$\cdots \longrightarrow 0 \longrightarrow 0 \longrightarrow P_o \xrightarrow{\varepsilon} P \longrightarrow 0$$

is a projective resolution, where $P_o = P$ and $\varepsilon = 1_P$.

We can dispense with the phrase "similarly in the other variable" after the following discussion.

DEFINITION. If R is a ring, then its **opposite ring**, R^{op}, is the ring having the same additive group as R but with multiplication defined by $a \circ b = ba$.

If R is commutative, then $R = R^{op}$.

Exercises 8.1. If A is a left R-module, then A is a right R^{op}-module; Similarly, right R-modules are left R^{op}-modules.

8.2. Given the situation A_R and ${}_RB$, prove that $\tau: A \otimes_R B \to B \otimes_{R^{op}} A$ defined by $a \otimes b \mapsto b \otimes a$ is an isomorphism. In particular, if R is commutative (and $ra = ar$, $rb = br$), then $A \otimes_R B \cong B \otimes_R A$.

8.3. If R and S are rings, prove that $R \otimes_Z S$ is a ring if we define multiplication by $(r \otimes s)(r' \otimes s') = rr' \otimes ss'$.

8.4. Let Λ denote the ring $R \otimes_Z S^{op}$, where R, S are rings. Prove that every 2-sided module ${}_RA_S$ is a left Λ-module when one defines $(r \otimes s)a = ras$.

8.5. If P is a projective left R-module, then P is also projective as a right R^{op}-module.

THEOREM 8.5. $\text{Tor}_n^R(A, B) \cong \text{Tor}_n^{R^{op}}(B, A)$, *all* n, *all* A, B.

Proof: If $\mathbf{P}_A = \cdots \to P_1 \to P_o \to O$, then the maps

$$\tau: P_n \otimes B \to B \otimes P_n$$

of exercise 8.2 define a chain map $\tau: \mathbf{P}_A \otimes B \to B \otimes \mathbf{P}_A$. Moreover, τ is an isomorphism (exercise 6.1), so that

$$H_n(\mathbf{P}_A \otimes B) \cong H_n(B \otimes \mathbf{P}_A).$$

Our result follows from Theorem 6.16.

COROLLARY 8.6. *If* R *is commutative,*

$$\text{Tor}_n^R(A, B) \cong \text{Tor}_n^R(B, A)$$

for all n, *all* A, B.

Clearly there is no analogue of this for Ext.

We now improve Theorem 8.4.

THEOREM 8.7. *If* F *is flat, then* $\text{Tor}_n(F, B) = O$ *for all* B *and all* $n > O$.

Proof: We do an induction on n. There is an exact sequence

$$O \longrightarrow K \xrightarrow{i} P \longrightarrow B \longrightarrow O$$

where P is projective. There arises an exact sequence

$$\text{Tor}_1(F, P) \longrightarrow \text{Tor}_1(F, B) \longrightarrow F \otimes K \xrightarrow{1 \otimes i} F \otimes P.$$

The first Tor $= O$ since P is projective. Since F is flat, $1 \otimes i$ is monic, so that $\text{Tor}_1(F,B) = O$. For the inductive step, we look further out in the exact sequence:

$$\text{Tor}_{n+1}(F,P) \longrightarrow \text{Tor}_{n+1}(F,B) \longrightarrow \text{Tor}_n(F,K) \quad .$$

The two outside terms are O, by Theorem 8.4 and induction, so that $\text{Tor}_{n+1}(F,B) = O$.

Easy induction arguments are common here; indeed, they are so common, there is a name for the technique. Suppose $O \to A' \to A \to A'' \to O$ is exact and A is well-behaved in the sense that, for $n > O$, all homology groups of A vanish (e.g., projective, injective, flat). An induction argument based on the exact triangle applied to such a sequence is called **dimension shifting.**

THEOREM 8.8. *If* $\text{Tor}_1(F,B) = O$ *for all* B, *then* F *is flat.*

Proof: If $O \to B' \xrightarrow{i} B \to B'' \to O$ is exact, then so is

$$\text{Tor}_1(F,B'') \longrightarrow F \otimes B' \xrightarrow{1 \otimes i} F \otimes B \quad .$$

Since $\text{Tor}_1(F,B'') = O$, $1 \otimes i$ is monic and F is flat.

Exercises 8.6. If $O \to A' \to A \to A'' \to O$ is exact and A' and A'' are flat, then A is flat.

8.7. If A'' is flat, then the exact sequence $O \to A' \to A \to A'' \to O$ is pure exact.

THEOREM 8.9. $\text{Tor}_n(\Sigma_k A_k, B) \cong \Sigma_k \text{Tor}_n(A_k, B)$.

Proof: Similar to the proof of Theorem 7.10, i.e., dimension shifting.

Exercise 8.8. Prove that $\text{Tor}_n(\varinjlim A_k, B) \cong \varinjlim \text{Tor}_n(A_k, B)$.

THEOREM 8.10. *If R is commutative,* $\operatorname{Tor}_n^R(A, B)$ *is an R-module.*

Proof: In this case, the complex defining Tor has as its terms R-modules and R-maps.

THEOREM 8.11. *If R is commutative and $\mu : A \to A$ is multiplication by $r \in R$, then so is* $\mu_*: \operatorname{Tor}_n(A, B) \to \operatorname{Tor}_n(A, B)$.

Proof: Similar to the proof of Theorem 7.13.

We now inquire about the etymology of Tor. If R is a domain, the notion of torsion is sensible: $tA = \{a \in A : ra = O$ for some $r \in R,\ r \neq O\}$ is a submodule of A. If R is not a domain, tA need not be a submodule: if $ra = O$ and $r'a' = O$, then $rr'(a + a') = O$, but there is no guarantee that $rr' \neq O$.

For a while, we assume R is a domain having quotient field Q; we denote the module Q/R by K.

LEMMA 8.12. *If A is a torsion R-module, then* $\operatorname{Tor}_1(K, A) \cong A$.

Proof: Exactness of $O \to R \to Q \to K \to O$ gives exactness of

$$\operatorname{Tor}_1(Q, A) \longrightarrow \operatorname{Tor}_1(K, A) \longrightarrow R \otimes A \longrightarrow Q \otimes A \ .$$

The first term is O since Q is flat (Corollary 3.32), while the last term is O since Q is divisible and A is torsion.

LEMMA 8.13. *For any R-module A,* $\operatorname{Tor}_n(K, A) = O$ *for $n \geq 2$.*

Proof: We have exactness of

$$\operatorname{Tor}_n(Q, A) \longrightarrow \operatorname{Tor}_n(K, A) \longrightarrow \operatorname{Tor}_{n-1}(R, A) \ ,$$

and the outside terms are O since Q, R are flat and $n - 1 \geq 1$.

LEMMA 8.14. *If A is torsion-free, then* $\operatorname{Tor}_1(K, A) = O$.

Proof: We saw in Theorem 4.21 that A can be imbedded in a torsion-free injective module E. Since E is just a vector space over Q, it is a sum of copies of Q, hence is flat. Exactness of $O \to A \to E \to X \to O$ gives exactness of

$$\mathrm{Tor}_2(K, X) \longrightarrow \mathrm{Tor}_1(K, A) \longrightarrow \mathrm{Tor}_1(K, E) .$$

The first term is O, by Lemma 8.13; the last term is O since E is flat.

THEOREM 8.15. *For every module A, $\mathrm{Tor}_1(K, A) \cong tA$, the torsion submodule of A.*

Proof: Exactness of $O \to tA \to A \to A/tA \to O$ gives exactness of

$$\mathrm{Tor}_2(K, A/tA) \longrightarrow \mathrm{Tor}_1(K, tA) \longrightarrow \mathrm{Tor}_1(K, A) \longrightarrow \mathrm{Tor}_1(K, A/tA) .$$

The first term is O, by Lemma 8.13; the last is O by Lemma 8.14.

Exercise 8.9. The torsion submodule t is a functor: on objects A, its value is tA; if $f\colon A \to A'$, then $tf = f \mid tA$. Prove that t is naturally equivalent to $\mathrm{Tor}_1(K, \ \ )$.

COROLLARY 8.16. *For every module A, there is an exact sequence*

$$O \longrightarrow tA \longrightarrow A \longrightarrow Q \otimes A \longrightarrow K \otimes A \longrightarrow O .$$

COROLLARY 8.17. *A module A is torsion if and only if $Q \otimes A = O$.*

Proof: We have known necessity for a long time. For the converse, use the exact sequence of Corollary 8.16.

LEMMA 8.18. *If B is torsion, then $\mathrm{Tor}_n(A, B)$ is torsion, all A, all $n \geq O$.*

Proof: We do a dimension shift.

If $n = O$, then each generator $a \otimes b$ is torsion, so that $A \otimes B$ is torsion.

If $n = 1$, there is an exact sequence $O \to N \to P \to A \to O$, where P is projective, and this gives exactness of

$$O = \mathrm{Tor}_1(P, B) \longrightarrow \mathrm{Tor}_1(A, B) \longrightarrow N \otimes B \ .$$

Since $N \otimes B$ is torsion, by the case $n = O$, so is its submodule $\mathrm{Tor}_1(A, B)$.

For the inductive step, look further out in the sequence. We have exactness of

$$O = \mathrm{Tor}_{n+1}(P, B) \longrightarrow \mathrm{Tor}_{n+1}(A, B) \longrightarrow \mathrm{Tor}_n(N, B) \longrightarrow \mathrm{Tor}_n(P, B) = O.$$

Since $\mathrm{Tor}_n(N, B)$ is torsion, by induction, so is $\mathrm{Tor}_{n+1}(A, B)$.

THEOREM 8.19. *For all* A, B *and all* $n \geq 1$, $\mathrm{Tor}_n(A, B)$ *is torsion.*

Proof: Suppose $n = 1$. We consider first the special case when B is torsion-free. As in Theorem 4.21, there is an exact sequence

$$O \longrightarrow B \longrightarrow E \longrightarrow X \longrightarrow O$$

where E is a vector space over Q. If we replace E by the subspace spanned by B, we may assume that X is torsion. There results an exact sequence

$$\mathrm{Tor}_2(A, X) \longrightarrow \mathrm{Tor}_1(A, B) \longrightarrow \mathrm{Tor}_1(A, E) \ .$$

Now $\mathrm{Tor}_2(A, X)$ is torsion, by Lemma 8.18, while $\mathrm{Tor}_1(A, E) = O$ since E is flat (it is a sum of copies of Q). Thus $\mathrm{Tor}_1(A, B)$ is a quotient of a torsion module, hence is torsion. Now let B be any module. Exactness of $O \to tB \to B \to B/tB \to O$ gives exactness of

$$\mathrm{Tor}_1(A, tB) \longrightarrow \mathrm{Tor}_1(A, B) \longrightarrow \mathrm{Tor}_1(A, B/tB) .$$

The flanking terms are torsion, for tB is torsion and B/tB is torsion-free; therefore $\mathrm{Tor}_1(A, B)$ is torsion. The proof is completed by dimension shifting.

This last result does not hold for $n = O$, i.e., $A \otimes B$ need not be torsion. It may happen, however, that A and B are torsion-free, but $A \otimes B$ is not torsion-free. For example, if $R = k[x, y]$, where k is a field, and if I is the ideal (x, y), then I is torsion-free but $I \otimes_R I$ is not torsion-free.

Exercises 8.10. If R is commutative, and if A and B are flat R-modules, then $A \otimes B$ is flat.

8.11. If R is commutative and I, J are ideals, then

$$\mathrm{Tor}_2(R/I, R/J) \cong \ker(I \otimes J \to IJ)$$

(the map is multiplication $i \otimes j \mapsto ij$).

8.12. If R is commutative, I, J are ideals, and $n > 2$, then

$$\mathrm{Tor}_n(R/I, R/J) \cong \mathrm{Tor}_{n-2}(I, J) .$$

The following is true, but harder: with the same hypotheses as exercise 8.11, $\mathrm{Tor}_1(R/I, R/J) \cong (I \cap J)/IJ$.

8.13. If A, B are finite abelian groups, then

$$A \otimes B \cong \mathrm{Tor}_1(A, B) .$$

8.14. Prove that $\mathrm{Tor}_1^Z(A, Z/nZ) \cong A_n = \{a \in A : na = O\}$.

The following results hold in more generality than stated here, and the reader is referred to MacLane for the not too difficult proofs. We examine the extent that homology commutes with $\otimes$ and Hom.

KÜNNETH FORMULAS

Let $\mathbf{A}$, $\mathbf{C}$ be complexes of abelian groups, and let each A_n in $\mathbf{A}$ be torsion-free (= flat). There is a split exact sequence

$$O \longrightarrow \sum_{i+j=n} H_i(\mathbf{A}) \otimes H_j(\mathbf{C}) \longrightarrow H_n(\mathbf{A} \otimes \mathbf{C})$$

$$\longrightarrow \sum_{i+j=n-1} \mathrm{Tor}_1^Z(H_i(\mathbf{A}), H_j(\mathbf{C})) \longrightarrow O .$$

Let $\mathbf{A}$, $\mathbf{C}$ be complexes of abelian groups, and suppose each A_n in $\mathbf{A}$ is free (= projective). There is a split exact sequence

$$O \longrightarrow \prod_{i+j=n-1} \mathrm{Ext}_Z^1(H_i(\mathbf{A}), H_j(\mathbf{C})) \longrightarrow H_n(\mathrm{Hom}(\mathbf{A}, \mathbf{C}))$$

$$\longrightarrow \prod_{i+j=n} \mathrm{Hom}(H_i(\mathbf{A}), H_j(\mathbf{C})) \longrightarrow O .$$

If one looks at the special case when $\mathbf{C}$ is an abelian group C concentrated in degree O, he obtains

UNIVERSAL COEFFICIENT THEOREMS

If $\mathbf{A}$ is a complex of torsion-free abelian groups, then for each group C and each n, there is a split exact sequence

$$O \longrightarrow H_n(\mathbf{A}) \otimes C \longrightarrow H_n(\mathbf{A} \otimes C) \longrightarrow \mathrm{Tor}_1^Z(H_{n-1}(\mathbf{A}), C) \longrightarrow O .$$

If $\mathbf{A}$ is a complex of free abelian groups, then, for each group C and each n, there is a split exact sequence

$$O \quad \mathrm{Ext}^1_Z(H_{n-1}(\mathbf{A}), C) \longrightarrow H_n(\mathrm{Hom}(\mathbf{A}, C)) \longrightarrow \mathrm{Hom}(H_n(\mathbf{A}), C) \longrightarrow O.$$

Exercises 8.15. If B is a torsion-free group, prove that

$$\mathrm{Tor}^Z_1(A, B \otimes C) \cong \mathrm{Tor}^Z_1(A \otimes B, C) .$$

8.16. If B is a free abelian group, prove that

$$\mathrm{Ext}^1_Z(A \otimes B, C) \cong \mathrm{Ext}^1_Z(A, \mathrm{Hom}(B, C)) .$$

For the etymology of Universal Coefficients, we return to §1 and topology. Given a space X, its homology groups $H_n(X)$ are constructed as the homology groups of the complex of free abelian groups

$$\mathbf{A} = \cdots \longrightarrow S_{n+1}(X) \longrightarrow S_n(X) \longrightarrow \cdots .$$

The modified version, $H_n(X; G)$, is homology "with coefficients in G," and is obtained as the homology of $\mathbf{A} \otimes G$. The first universal coefficient theorem thus allows us to compute $H_n(X; G)$ if we know $H_n(X) = H_n(X; Z)$:

$$H_n(X; G) \cong H_n(X) \otimes G \oplus \mathrm{Tor}^Z_1(H_{n-1}(X), G) .$$

Similarly, the second universal coefficient theorem shows that the homology groups of X determine the cohomology groups:

$$H^n(X; G) \cong \mathrm{Hom}(H_n(X), G) \oplus \mathrm{Ext}^1_Z(H_{n-1}(X), G) .$$

§9. *Son of Particular Rings*

In this chapter, we address the problem of the interrelationships between a ring R and Ext and Tor. We begin, however, with a minor illustration that these functors are good for something, and so we use Ext to prove a theorem (9.3) that does not mention Ext in either its hypothesis or conclusion.

THEOREM 9.1. *If R is a left hereditary, then* $\mathrm{Ext}_R^n(A,B) = O$ *for all left R-modules A, B, all $n \geq 2$.*

Proof: There is an exact sequence

$$O \longrightarrow P_1 \longrightarrow P_o \longrightarrow A \longrightarrow O ,$$

where P_o is projective. By Corollary 4.5, P_1 is also projective, so that the short exact sequence above is a projective resolution of A. The result follows.

THEOREM 9.2. *If R is a Dedekind ring and A is a torsion-free R-module, then, for every R-module B,* $\mathrm{Ext}_R^1(A, B)$ *is a divisible R-module.*

Proof: There is an exact sequence

$$O \longrightarrow A \longrightarrow E \longrightarrow E'' \longrightarrow O ,$$

where E is torsion-free divisible (Corollary 8.16), and this gives rise to an exact sequence

$$\mathrm{Ext}^1(E, B) \longrightarrow \mathrm{Ext}^1(A, B) \longrightarrow \mathrm{Ext}^2(E'', B) .$$

The last term is O, by Theorem 9.1, so that the first map is epic. But $\mathrm{Ext}^1(E, B)$ is divisible, by exercise 7.4, so that its image $\mathrm{Ext}^1(A, B)$ is also divisible.

If R is a domain, B an R-module, then its torsion submodule need not be a summand of B. For example, if $R = Z$ and $B = \Pi_p Z/pZ$ (the product is over all primes p), then one may check that $tB = \Sigma_p Z/pZ$ and that B/tB is divisible. Since B has no nonzero divisible subgroups, we cannot have $B \cong tB \oplus B/tB$.

THEOREM 9.3. *Let R be a Dedekind ring, and let B be an R-module with $r(tB) = O$ for some nonzero $r \in R$. Then tB is a summand of B.*

Proof: We must show that $O \to tB \to B \to B/tB \to O$ is split. Since B/tB is torsion-free, it will suffice to prove $\mathrm{Ext}^1(A, T) = O$ whenever A is torsion-free and $T \cong tB$. Now $\mathrm{Ext}^1(A, T)$ is divisible, by Theorem 9.2. Since $rT = O$, we have $r\mathrm{Ext}^1(A, T) = O$, by exercise 7.5. Hence $\mathrm{Ext}^1(A, T) = O$, for divisibility implies that multiplication by r is epic.

The next result is useful.

THEOREM 9.4. *Let R be a commutative noetherian ring, and let A, B be f.g. R-modules. Then, for all n, $\mathrm{Ext}^n(A, B)$ and $\mathrm{Tor}_n(A, B)$ are f.g. R-modules.*

Proof: Observe first that the theorem is true when $n = O$, i.e., $\mathrm{Hom}(A, B)$ and $A \otimes B$ are f.g. R-modules. For general n, we build a good projective resolution of A. Since A is f.g., there is an exact sequence

$$O \longrightarrow K_o \longrightarrow P_o \longrightarrow A \longrightarrow O\ ,$$

where P_o is f.g. projective. Since R is noetherian, Theorem 4.25 gives K_o f.g.; there is thus an exact sequence

$$O \longrightarrow K_1 \longrightarrow P_1 \longrightarrow K_o \longrightarrow O ,$$

where P_1 is f.g. projective. The usual iteration gives a projective resolution $\cdots \to P_1 \to P_o \to A \to O$ of A in which each P_n is f.g.

Now if B is f.g., each $\mathrm{Hom}(P_n, B)$ is f.g., hence its submodule $\ker d^*_{n+1}$ is also f.g. (Theorem 4.25 again), and, finally, $\ker d^*_{n+1}/\mathrm{im}\ d^*_n = \mathrm{Ext}^n(A, B)$ is f.g. A similar argument works for Tor when we replace Hom by $\otimes$.

We now measure how far a module is from being projective.

DEFINITION. Let R be a ring, and A a left R-module. We say $hd(A) \leq n$ (hd abbreviates **homological dimension**) if there is a projective resolution

$$O \longrightarrow P_n \longrightarrow \cdots \longrightarrow P_1 \longrightarrow P_o \longrightarrow A \longrightarrow O .$$

We say $hd(A) = n$ if $hd(A) \leq n$ but $hd(A) \not\leq n-1$; in words, there is a projective resolution of length n, but no shorter one. Finally, $hd(A) = \infty$ if there is no n with $hd(A) \leq n$.

EXAMPLES. 1. $hd(A) = O$ if and only if A is projective.

2. If R is left hereditary, $hd(A) \leq 1$ for every R-module A (this is in the proof of Theorem 9.1).

DEFINITION. If $\cdots \to P_1 \xrightarrow{d_1} P_o \xrightarrow{\varepsilon} A \to O$ is a projective resolution, its n-th **kernel** is $\ker d_n$, where $n \geq 1$. The O-th **kernel** is $\ker \varepsilon = \mathrm{im}\ d_1$.

Exercise 9.1. Define two R-modules A, B to be **projectively equivalent** if there are projective modules P, Q with $A \oplus P \cong B \oplus Q$. (This is an equivalence relation.) Prove that, for each n, the n-th kernels of two projective resolutions of A are projectively equivalent.

LEMMA 9.5. *Let* $\cdots \to P_1 \to P_o \to A \to O$ *be a projective resolution of* A *with n-th kernel* K_n. *Then, for all modules* B,

$$\operatorname{Ext}^{n+1}(A, B) \cong \operatorname{Ext}^1(K_{n-1}, B).$$

Proof: First, we claim that for all $n \geq 1$, all B,

$$\operatorname{Ext}^{n+1}(A, B) \cong \operatorname{Ext}^n(K_o, B).$$

This follows from the fact that we may compute $\operatorname{Ext}^{n+1}(A, B)$ using the given resolution of A, while $\operatorname{Ext}^n(K_o, B)$ may be computed from the projective resolution of K_o:

$$\cdots \longrightarrow P_2 \longrightarrow P_1 \longrightarrow K_o \longrightarrow O.$$

The Lemma now follows by iteration of this isomorphism:

$$\begin{aligned}\operatorname{Ext}^{n+1}(A, B) &\cong \operatorname{Ext}^n(K_o, B)\\ &\cong \operatorname{Ext}^{n-1}(K_1, B) \cong \cdots \cong \operatorname{Ext}^1(K_{n-1}, B).\end{aligned}$$

THEOREM 9.6. *The following are equivalent for a left R-module* A:

i) $hd(A) \leq n$;

ii) $\operatorname{Ext}^k(A, B) = O$ *for all modules* B *and all* $k \geq n+1$;

iii) $\operatorname{Ext}^{n+1}(A, B) = O$ *for all modules* B;

iv) *every projective resolution of* A *has a projective* $(n-1)$*st kernel.*

Proof: i) $\Longrightarrow$ ii). There is a projective resolution of A with $P_k = O$ for all $k \geq n+1$. Therefore $\operatorname{Hom}(P_k, B) = O$ for all $k \geq n+1$ and so $\operatorname{Ext}^k(A, B) = O$ for all $k \geq n+1$.

ii) $\Longrightarrow$ iii). Trivial.

iii) $\Longrightarrow$ iv). Take a projective resolution of A with $(n-1)$ st kernel K_{n-1}. Lemma 9.5 gives $\mathrm{Ext}^{n+1}(A, B) \cong \mathrm{Ext}^1(K_{n-1}, B)$, so that $\mathrm{Ext}^1(K_{n-1}, B) = O$ for all B; therefore K_{n-1} is projective, by Corollary 7.9.

iv) $\Longrightarrow$ i). If $\cdots \to P_n \to P_{n-1} \xrightarrow{d_{n-1}} \cdots \to P_o \to A \to O$ is a projective resolution of A, so is

$$O \longrightarrow K_{n-1} \longrightarrow P_{n-1} \longrightarrow \cdots \longrightarrow P_o \longrightarrow A \longrightarrow O\,;$$

therefore, $hd(A) \leq n$.

DEFINITION. If R is a ring, $\ell pD(R)$, its **left projective global dimension,** is defined by

$$\ell pD(R) = \sup\{hd(A)\colon\ A \in {}_R\mathfrak{M}\}\,.$$

COROLLARY 9.7. *$\ell pD(R) \leq n$ if and only if $\mathrm{Ext}^{n+1}(A, B) = O$ for all left R-modules A and B.*

Proof: Immediate from Theorem 9.6 (iii).

Exercises 9.2. $\ell pD(R) = O$ if and only if R is semisimple.

9.3. Let R be a quasi-Frobenius ring that is not semisimple. Prove that $\ell pD(R) = \infty$.

THEOREM 9.8. *$\ell pD(R) \leq 1$ if and only if R is left hereditary.*

Proof: Sufficiency is Theorem 9.1. For the converse, suppose $\ell pD(R) \leq 1$. Suppose P is projective and P' is a submodule of P. There is thus an exact sequence

$$O \longrightarrow P' \longrightarrow P \longrightarrow A \longrightarrow O\,,$$

where $A = P/P'$. In the usual way, we may construct a projective resolution of A whose O-th kernel is P'. By hypothesis, $hd(A) \leq 1$, so that P' is projective, by Theorem 9.6 (iv). Hence R is left hereditary, by Theorem 4.10.

All of this can be repeated using injective modules instead of projectives. We need only state the definitions and results.

DEFINITION. Let R be a ring and B a left R-module. We say $id(B) \leq n$ (id appreviates **injective dimension**) if there is an injective resolution

$$O \longrightarrow B \longrightarrow E_o \longrightarrow \cdots \longrightarrow E_n \longrightarrow O \ .$$

We say $id(B) = n$ if $id(B) \leq n$ but $id(B) \not\leq n-1$; finally $id(B) = \infty$ if there is no n with $id(B) \leq n$.

DEFINITION. If $O \to B \overset{\varepsilon}{\to} E_o \xrightarrow{d_o} E_1 \to \cdots$ is an injective resolution, its n-th **image** is im d_{n-1} (if $n \geq 1$); its O-th **image** is im ε = ker d_o .

LEMMA 9.9. *Let $O \to B \to E_o \to E_1 \to \cdots$ be an injective resolution of B with n-th image I_n. Then for all modules A,*

$$\text{Ext}^{n+1}(A, B) \cong \text{Ext}^1(A, I_{n-1}) \ .$$

THEOREM 9.10. *The following are equivalent for a left R-module B:*

i) $id(B) \leq n$;

ii) $\text{Ext}^k(A, B) = O$ *for all modules A and all $k \geq n+1$;*

iii) $\text{Ext}^{n+1}(A, B) = O$ *for all modules A;*

iv) *every injective resolution of B has an injective $(n-1)$ st image.*

DEFINITION. If R is a ring, $\ell iD(R)$, its **left injective global dimension**, is defined by

$$\ell iD(R) = \sup\{id(B): B \in {}_R\mathfrak{M}\} .$$

COROLLARY 9.11. *$\ell iD(R) \leq n$ if and only if* $\mathrm{Ext}^{n+1}(A, B) = 0$ *for all left R-modules A and B.*

THEOREM 9.12. *For any ring R, $\ell iD(R) = \ell pD(R)$.*

Proof: Corollaries 9.11 and 9.7.

In view of Theorem 9.12, one defines the **left global dimension** of R, $\ell D(R)$, as the common value of $\ell iD(R)$ and $\ell pD(R)$.

REMARK. If one considers right R-modules instead of left R-modules, he may define the **right global dimension** of R, denoted $rD(R)$. Theorem 4.3 (essentially the Wedderburn structure theorem) gives $\ell D(R) = 0$ if and only if $rD(R) = 0$. On the other hand, Kaplansky's example (see §4) of a ring R that is right hereditary but not left hereditary is an example when the two dimensions differ. In fact, there $rD(R) = 1$ and $\ell D(R) = 2$. Auslander (Nagoya M. J., 1955) proved that if R is both left and right noetherian, then $\ell D(R) = rD(R)$. Jensen (Math. Scand., 1966) proved that if all ideals are countably generated, then the left and right global dimensions differ by at most 1. Small (Proc. N.A.S., 1966) gave an example of a ring R with $rD(R) = 1$ and $\ell D(R) = 3$. In 1969, Jategaonkar gave an example of a left noetherian ring whose left global dimension is m and whose right global dimension is n, where $0 < m \leq n \leq \infty$.

The next problem is the computation of the global dimension.

LEMMA 9.13. *A left R-module B is injective if and only if* $\mathrm{Ext}^1(R/I, B) = O$ *for all left ideals I of R.*

Proof: If one applies Hom(, B) to the exact sequence $O \to I \to R \to R/I \to O$, he sees that the theorem follows immediately from Theorem 3.19.

THEOREM 9.14. (Auslander) *For any ring R,*

$$\ell D(R) = \sup\{hd(R/I):\ I = \textit{left ideal of } R\}\ .$$

Proof: If $\sup\{hd(R/I)\} = \infty$, we are done; therefore, we may assume $hd(R/I) \leq n$ for all left ideals I, and we shall prove that $id(B) \leq n$ for every module B. The theorem will then follow from Theorem 9.12. Thus, assuming $\mathrm{Ext}^{n+1}(R/I, B) = O$ for every ideal I, we must show that $\mathrm{Ext}^{n+1}(A, B) = O$ for every A. Take an injective resolution of B having n-th image I_n. If $\mathrm{Ext}^{n+1}(R/I, B) = O$, then $\mathrm{Ext}^1(R/I, I_{n-1}) = O$ (Lemma 9.9), and so I_{n-1} is injective, by Lemma 9.13. Hence, for every module A, $\mathrm{Ext}^1(A, I_{n-1}) = O$ and, using Lemma 9.9 again, $\mathrm{Ext}^{n+1}(A, B) = O$.

Thus, to compute the global dimension of a ring, it suffices to know the dimensions of the cyclic modules.

Exercises 9.4. Let there be given a projective resolution of A with n-th kernel K_n and an injective resolution of B with n-th image I_n. For all $n \geq 1$, prove that $\mathrm{Ext}^1(K_{n-1}, B) \cong \mathrm{Ext}^1(A, I_{n-1})$.

9.5. Let $O \to A' \to A \to A'' \to O$ be exact, where A is projective. If A'' is not projective, prove that

$$hd(A'') = hd(A') + 1\ .$$

(We agree that $\infty + 1 = \infty$.)

9.6. Given a family of modules $\{A_k: k \in K\}$, then

$$hd(\Sigma A_k) = \sup\{hd(A_k)\} .$$

Conclude that if $\ell D(R) = \infty$, there exists a left R-module A with $hd(A) = \infty$.

9.7. If $O \to A' \to A \to A'' \to O$ is exact and two of the modules have finite homological dimension, so does the third.

9.8. If $O \to A' \to A \to A'' \to O$ is exact, then

$$hd(A) \leq \max\{hd(A'), hd(A'')\} ;$$

moreover, there is equality unless the sequence doesn't split and $hd(A'') = hd(A') + 1$.

9.9. If $O \to A' \to A \to A'' \to O$ is exact, then

$$hd(A'') \leq 1 + \max\{hd(A'), hd(A)\} .$$

REMARK. The question arises whether there is an analogue of Lemma 9.13 to test for projectivity. At any rate, the obvious candidates do not seem to work. If we assume $\text{Ext}_R^1(P, R/I) = O$ for all I, this only gives P torsion-free (at least when R is Dedekind, by Theorem 9.3). If we assume that $\text{Ext}_R^1(P, I) = O$ for all I, the question whether P is projective is open. In particular, if $R = Z$, then every nonzero ideal I is isomorphic to Z (I is principal), and the question is whether $\text{Ext}^1(P, Z) = O$ implies P free; this is called *Whitehead's problem.* It is known that if P is countable, then P is free.

Exercises. 9.10. Let R be a Dedekind ring, and let $K = Q/R$, where Q is the quotient field of R. Prove that an R-module A is injective if and only if $\text{Ext}_R^1(K, A) = O$.

9.11. If R is a Dedekind ring, prove that an R-module A is projective if and only if $\operatorname{Ext}_R^1(A, P) = O$ for all projective modules P.

There are other dimensions associated to modules (hence to rings, by taking the sup of the dimension of modules).

9.11A Define a **flat resolution** of a left R-module B as an exact sequence

$$\cdots \longrightarrow X_1 \longrightarrow X_o \longrightarrow B \longrightarrow O$$

in which each X_i is flat. For every right R-module A, prove that $H_n(A \otimes \mathrm{X}_B) \cong \operatorname{Tor}_n^R(A, B)$, all n. Hint: Dimension shifting.

DEFINITION. A right R-module A has **flat dimension** $\leq n$, $fd(A) \leq n$, if there is an exact sequence

$$O \longrightarrow F_n \longrightarrow P_{n-1} \longrightarrow \cdots \longrightarrow P_o \longrightarrow A \longrightarrow O,$$

where F_n is flat and the P_i are projective; $fd(A) = n$ if $fd(A) \leq n$ and there is no shorter sequence of the above form; $fd(A) = \infty$ if there is no n with $fd(A) \leq n$.

Exercises 9.12. If K_n is the n-th kernel of a projective resolution of A, then $\operatorname{Tor}_{n+1}(A, B) \cong \operatorname{Tor}_1(K_{n-1}, B)$.

9.13. The following are equivalent for a right R-module A:

i) $fd(A) \leq n$;

ii) every projective resolution of A has a flat $(n{-}1)$ st kernel;

iii) $\operatorname{Tor}_k(A, B) = O$ for all $k \geq n+1$, all left R-modules B;

iv) $\operatorname{Tor}_{n+1}(A, B) = O$ for all left R-modules B.

DEFINITION. The **right weak dimension** of R, $rwD(R)$, is defined by

$$rwD(R) = \sup\{fd(A) : A \in \mathfrak{M}_R\}.$$

THEOREM 9.15. *$rwD(R) \leq n$ if and only if $\mathrm{Tor}_{n+1}(A, B) = O$ for all right R-modules A and all left R-modules B.*

Proof: Immediate from exercise 9.13.

If one considers left R-modules, he may define the **left weak dimension of** R, $\ell wD(R)$.

THEOREM 9.16. *For any ring R, $rwD(R) = \ell wD(R)$.*

Proof: One may prove the left version of exercise 9.13, obtaining as corollary: $\ell wD(R) \leq n$ if and only if $\mathrm{Tor}_{n+1}(A, B) = O$ for all right R-modules A and all left R-modules B. The theorem follows by comparing this criterion with Theorem 9.15.

DEFINITION. The **weak dimension** of R, $wD(R)$, is the common value of $rwD(R)$ and $\ell wD(R)$.

The next result explains why this dimension is called 'weak.'

THEOREM 9.17. *For any ring R, $wD(R) \leq \min\{\ell D(R), rD(R)\}$.*

Proof: Given a left R-module A, we claim $fd(A) \leq hd(A)$: if a projective resolution of A has a projective $(n-1)$ st kernel, it has a flat $(n-1)$ st kernel, since projectives are flat. Hence $wD(R) \leq \ell D(R)$. A similar argument gives $wD(R) \leq rD(R)$.

It follows, for example, that $\mathrm{Tor}_2^R(A, B) = O$ for all A, B when R is a hereditary ring.

To compute $wD(R)$, it suffices to know the cyclic modules.

THEOREM 9.18. *For any ring R,*

$$wD(R) = \sup\{fd(R/I): I = \text{right ideal of } R\}$$
$$= \sup\{fd(R/J): J = \text{left ideal of } R\}.$$

Proof: First observe that a left R-module B is flat if and only if $\mathrm{Tor}_1(R/I, B) = O$ for all right ideals I; this follows from Theorem 3.36. The proof continues as the proof of Theorem 9.14 to give the first equality. The second equality is proved the same way, but now fix the first variable of Tor_1.

For noetherian rings, the weak dimension is nothing new.

THEOREM 9.19. *If R is right noetherian, then*

$$wD(R) = rD(R) \ .$$

Similarly, if R is left noetherian, $wD(R) = \ell D(R)$.

Proof: We claim that if A is a f.g. right R-module, then $fd(A) = hd(A)$. This will suffice, for we know that we may even compute $wD(R)$ and $rD(R)$ using only cyclic modules. Now $fd(A) \leq hd(A)$ always, so it is enough to prove $fd(A) \geq hd(A)$, i.e., if $\mathrm{Tor}_{n+1}(A, B) = O$ for all left R-modules B, then $\mathrm{Ext}^{n+1}(A, C) = O$ for all right R-modules C.

For any projective resolution of A, we have $O = \mathrm{Tor}_{n+1}(A, B) \cong \mathrm{Tor}_1(K_{n-1}, B)$ all B, where K_{n-1} is the $(n-1)$ st kernel, i.e., K_{n-1} is flat. Since R is right noetherian, there is a projective resolution of A for which every n-th kernel is f.g. (see proof of Theorem 9.4). It follows from Theorem 4.26 that K_{n-1} is projective. Therefore $\mathrm{Ext}^1(K_{n-1}, C) = O$ for all C, hence $\mathrm{Ext}^{n+1}(A, C) = O$, by Lemma 9.5.

COROLLARY 9.20. (Auslander) *If R is left and right noetherian, then $\ell D(R) = rD(R)$.*

Proof: Here, both these global dimensions equal $wD(R)$.

Our next aim is to give examples of rings of finite global dimension; in particular, we will examine polynomial rings. (Our account is stolen from Kaplansky, *Commutative Rings*.)

Suppose $\phi : R \to R^*$ is a ring homomorphism. Every R^*-module A^* can be made into an R-module (by defining $ra^* = \phi(r)a^*$) and every R^*-map is an R-map. In particular, R^* itself can be considered as an R-module. Each R^*-module A^* thus has two homological dimensions, which we denote $hd_R(A^*)$ and $hd_{R^*}(A^*)$.

Exercises 9.14. A ring map $\phi : R \to R^*$ defines an exact functor ${}_{R^*}\mathfrak{M} \to {}_R\mathfrak{M}$.

9.15. Let R be commutative, $x \in R$, x not a zero divisor. If P is a projective R-module, prove that $xP \neq O$.

THEOREM 9.21. *Suppose $\phi : R \to R^*$ is a ring map and A^* is an R^*-module. Then*

$$hd_R(A^*) \leq hd_{R^*}(A^*) + hd_R(R^*) \ .$$

Moreover, if there is equality for all nonzero A^* with $hd_{R^*}(A^*) \leq 1$, there is equality for all nonzero A^* with $hd_{R^*}(A^*) < \infty$.

Proof: If $hd_{R^*}(A^*) = \infty$, there is nothing to prove, so we may assume $hd_{R^*}(A^*) = n < \infty$; we do an induction on n. If $n = O$, then A^* is R^*-projective and $A^* \oplus B^* = F^*$, where F^* is R^*-free. It follows from exercise 9.6 that $hd_R(A^*) \leq hd_R(F^*) = hd_R(R^*)$.

Suppose $n > O$. There is an exact sequence of R^*-modules

$$O \longrightarrow K^* \longrightarrow F^* \longrightarrow A^* \longrightarrow O \ ,$$

where F^* is R^*-free. By exercise 9.9, we have

$$hd_R(A^*) \leq 1 + \max\{hd_R(K^*),\ hd_R(F^*)\}$$

$$= 1 + \max\{hd_R(K^*),\ hd_R(R^*)\}\ .$$

By exercise 9.5, $hd_{R*}(K^*) = n-1$, so that induction gives $hd_R(K^*) \leq n-1 + hd_R(R^*)$. Combining these inequalities gives

$$hd_R(A^*) \leq 1 + \max\{n-1 + hd_R(R^*),\ hd_R(R^*)\}$$

$$= n + hd_R(R^*)\ .$$

Finally, assume we have equality whenever $A^* \neq O$ and $hd_{R*}(A^*) \leq 1$; let $hd_R(R^*) = r$, and let $hd_{R*}(A^*) = n \geq 2$. By exercise 9.8, the exact sequence

$$O \longrightarrow K^* \longrightarrow F^* \longrightarrow A^* \longrightarrow O$$

gives $hd_{R*}(K^*) = n-1 \geq 1$; hence $K^* \neq O$. By induction, $hd_R(K^*) = n-1+r$. If $hd_R(A^*) = a$, then exercise 9.8 gives $r \leq \max\{n-1+r,\ a\}$. Now we cannot have equality here, for $\max\{n-1+r,\ a\} \geq n-1+r > r$; the rest of exercise 9.8 gives $hd_R(A^*) = hd_R(K^*) + 1 = (n-1+r) + 1 = n + r$.

If R is commutative, we write $D(R)$ instead of $\ell D(R)$.

THEOREM 9.22. *Let R be commutative, $x \in R$ not a unit and not a zero divisor, and $R^* = R/xR$. Then, if $D(R^*) < \infty$,*

$$D(R) \geq D(R^*) + 1\ .$$

Proof: It suffices to show that if A^* is a nonzero R^*-module with $hd_{R*}(A^*) < \infty$, then

$$hd_R(A^*) = hd_{R*}(A^*) + 1\ .$$

Since x is not a zero divisor, there is an exact sequence

$$O \longrightarrow R \xrightarrow{x} R \longrightarrow R^* \longrightarrow O\ ;$$

hence $hd_R(R^*) \leq 1$. In fact, $hd_R(R^*) = 1$, for if this sequence splits, xR is a summand of R and x would be a zero divisor. It follows from Theorem 9.21 that

$$hd_R(A^*) \leq hd_{R*}(A^*) + 1\,.$$

Moreover, if we have equality whenever $hd_{R*}(A^*) \leq 1$, we are done; it suffices to eliminate strict inequality.

If $hd_{R*}(A^*) = 0$, then $hd_R(A^*) \leq 1$ and strict inequality means $hd_R(A^*) = 0$, i.e., A^* is R-projective. This cannot happen, for $xA^* = 0$, while $xP \neq 0$ when P is a nonzero projective (exercise 9.15).

If $hd_{R^*}(A^*) = 1$, then $hd_R(A^*) \leq 2$ and strict inequality means $hd_R(A^*) \leq 1$. We have already ruled out $hd_R(A^*) = 0$; we claim $hd_R(A^*) \neq 1$. Otherwise, there is an exact sequence of R-modules

$$0 \longrightarrow K \longrightarrow F \longrightarrow A^* \longrightarrow 0\;,$$

where F is R-free, and K is R-projective. Since $xA^* = 0$, there is an exact sequence of R^*-modules

$$0 \longrightarrow K/xF \longrightarrow F/xF \longrightarrow A^* \longrightarrow 0\;.$$

Now F/xF is R^*-free, so that $hd_{R*}(A^*) = 1$ gives K/xF R^*-projective. But then the exact sequence

$$0 \longrightarrow xF/xK \longrightarrow K/xK \longrightarrow K/xF \longrightarrow 0$$

splits, so that $xF/xK \cong F/K \cong A^*$, and this contradicts $hd_{R*}(A^*) = 1$.

REMARK. It can happen that $D(R) < \infty$ and $D(R^*) = \infty$, so that the assumption $D(R^*) < \infty$ is essential to the inequality of the theorem. If $R = Z$, we know $D(R) = 1$; if $x = 4$, then $R^* = Z/4Z$ is quasi-Frobenius (Corollary 4.29) and not semisimple; by exercise 9.3, $D(R^*) = \infty$.

LEMMA 9.23. *Let R be a subring of the commutative ring T (so that T is an R-module), and suppose R is an R-module summand of T. Then*

$$D(R) \leq D(T) + hd_R(T) .$$

Proof: If A is an R-module, then $B = \text{Hom}_R(T, A)$ is a T-module and A is an R-module summand of B: since $T = R \oplus S$, $B = \text{Hom}_R(R, A) \oplus \text{Hom}_R(S, A)$. Now

$$hd_R(B) \leq hd_T(B) + hd_R(T) ,$$

by Theorem 9.21, and

$$hd_R(A) \leq hd_R(B),$$

since A is a summand of B. Taking sups gives

$$D(R) \leq D(T) + hd_R(T) .$$

REMARK. An example of the situation described in Lemma 9.23 is $T = R[x]$. As an R-module, $R[x]$ is free of rank $\aleph_0$ and so $hd_R(R[x]) = O$; therefore $D(R) \leq D(R[x])$.

DEFINITION. An R-module A is **faithfully flat** if it is flat and $A \otimes_R B = O$ implies $B = O$.

EXAMPLE. If R is a domain (not a field) with quotient field Q, then R is faithfully flat; Q is flat, but is not faithfully flat ($Q \otimes B = O$ whenever B is torsion).

Exercises. *9.16.* If R is a commutative ring, then $R[x]$ is a faithfully flat R-module.

9.17. If A is a projective R-module, then $R[x] \otimes A$ is a projective $R[x]$-module.

9.18. Denote $R[x] \otimes A$ by $A[x]$. Show that

$$hd_R(A) = hd_{R[x]}(A[x]) \ .$$

THEOREM 9.24. (Hilbert's Theorem on Syzygies). *If R is a commutative ring, then*

$$D(R[x]) = D(R) + 1 \ .$$

Proof: By the remark after Lemma 9.23, $D(R[x]) \geq D(R)$; in particular, if $D(R) = \infty$, then $D(R[x]) = \infty$. Therefore, we may assume $D(R) = n < \infty$. Now Theorem 9.22 gives $D(R[x]) \geq n+1$, so that we need only prove the reverse inequality: if M is an $R[x]$-module, then $hd(M) \leq n + 1$.

Given an $R[x]$-module M, there is an exact sequence

$$0 \longrightarrow K \longrightarrow M[x] \xrightarrow{\ e\ } M \longrightarrow 0 \ ,$$

where $M[x] = R[x] \otimes M$ and e sends $x^i \otimes m \longmapsto x^i m$. We claim that $M[x] \cong K$ as $R[x]$-modules. Define $\beta: M[x] \to K$ by $\Sigma\, x^i \otimes m_i \longmapsto \Sigma\, x^i(1 \otimes x - x \otimes 1)\, m_i$; clearly $\operatorname{im} \beta \subset K$ and β is an $R[x]$-map. To show β is an isomorphism we write the formula in more detail:

$$\sum_{i=0}^{k} x^i \otimes m_i \longmapsto 1 \otimes xm_0 + \sum_{i=1}^{k} x^i \otimes (xm_i - m_{i-1}) - x^{k+1} \otimes m_k \ .$$

If $\Sigma\, x^i \otimes m_i \in \ker \beta$, then

$$O = -m_k = xm_k - m_{k-1} = \cdots = xm_1 - m_o,$$

so that each $m_i = O$ and β is monic. If $\Sigma^k x^i \otimes v_i \in K$, then $\Sigma_{i=o}^k x^i v_i = O$; the equations

$$v_o = xm_o, \quad v_1 = xm_1 - m_o, \ldots$$

can be solved recursively to show that β is epic.

We now have

$$hd_{R[x]}(K) = hd_{R[x]}(M[x]) = hd_R(M) \le n,$$

by exercise 9.18 and the assumption $D(R) = n$. By exercise 9.9,

$$hd(M) \le 1 + \max\{hd(K), hd(M[x])\} \le n + 1,$$

as we wished to show.

COROLLARY 9.25. *If k is a field, then*

$$D(k[x_1, \ldots, x_n]) = n.$$

Proof: Induction on n, noting that $D(k) = O$.

We end this chapter with some remarks about *localization*. If R is a commutative ring, a subset S of R is *multiplicatively closed* if $O \notin S$ and $s_1, s_2 \in S$ implies $s_1 s_2 \in S$. The most popular example of such an S is the (set-theoretic) complement of a prime ideal of R. Given R and a multiplicative subset S, one may form the *localization* R_S (if S consists of non-zero-divisors, R_S consists of all fractions r/s, $s \in S$; if S does contain zero divisors, one must be a bit more careful). One can show R_S is an R-algebra that is flat as an R-module. Also, to explain the name, if S is the complement of a prime ideal P, then R_S is a local ring.

If R is a local ring (now assumed commutative, noetherian), with maximal ideal M, then M/M^2 is an R/M-module, i.e., a vector space; let its dimension be d. One may prove there is a longest chain of prime ideals $P_o > P_1 > \cdots > P_n$, and $n \leq d$. If $n = d$, the local ring R is called *regular*.

Serre gave a homological characterization of regular local rings:

THEOREM. *A local ring R is regular if and only if $D(R) < \infty$. Moreover, in this case, $D(R) = n = d$ (all three integers coincide).*

Since R_S is flat, an immediate corollary is:

If R is a regular local ring, then R_S is also regular.

It can be shown that regular local rings are domains. In 1959, Auslander and Buchsbaum proved (using global dimension) that every regular local ring is a unique factorization domain. We refer the reader to Kaplansky's notes for details.

§10. *Cohomology of Groups Returns.*

In view of the discussion in §5, the following definition is reasonable.

DEFINITION: Let G be a group, A a G-module, and Z the integers considered as a trivial G-module.

$$H^n(G, A) = \mathrm{Ext}^n_{ZG}(Z, A) ;$$

$$H_n(G, A) = \mathrm{Tor}^{ZG}_n(Z, A) .$$

The groups H^n are the **cohomology groups** of G; the groups H_n are the **homology groups** of G.

From now on, we use the notation H^n or H_n instead of Ext^n or Tor_n; in particular, the long exact sequences will be so written.

We recall some results from §5. $H^o(G, A) = \mathrm{Hom}_G(Z, A) = A^G$; $H_o(G, A) = Z \otimes_G A = A_G$. If A is G-trivial, then $A^G = A = A_G$. Also, we saw that $H^1(G, A) \cong \mathrm{stab}(G, A) = \mathrm{Der}(G, A)/P\,\mathrm{Der}(G, A)$.

THEOREM 10.1. *If $\mathcal{G}$ is the augmentation ideal of G, then*

$$H_1(G, Z) \cong \mathcal{G}/\mathcal{G}^2 ,$$

where Z is G-trivial (as usual).

Proof: The augmentation sequence $O \to \mathcal{G} \to ZG \xrightarrow{\varepsilon} Z \to O$ induces exactness of

$$H_1(G, ZG) \longrightarrow H_1(G, Z) \xrightarrow{\partial} H_o(G, \mathcal{G})$$
$$\longrightarrow H_o(G, ZG) \xrightarrow{\varepsilon_*} H_o(G, Z) \longrightarrow O .$$

Now $H_o(G, Z) \cong Z$, since Z is G-trivial. Also, $H_o(G, ZG) \cong Z$, for $H_o(G, ZG) = Z \otimes_{ZG} ZG$. But any endomorphism of Z is either O or monic; since ε_* is epic, $\varepsilon_* \neq O$, and so ε_* is monic.

Exactness of the sequence now gives ∂ epic. Finally, $H_1(G, ZG) = O$ since ZG is projective. Therefore $\partial : H_1(G, Z) \to H_o(G, \mathcal{G})$ is an isomorphism. By Theorem 5.15, if A is a G-module, $H_o(G, A) \cong A/\mathcal{G}A$; in particular, $H_o(G, \mathcal{G}) \cong \mathcal{G}/\mathcal{G}^2$.

THEOREM 10.2. *For any group* G, *the additive group* $\mathcal{G}/\mathcal{G}^2$ *is isomorphic to the multiplicative group* G/G', *where* G' *is the commutator subgroup of* G.

Proof: Define $\theta : G \to \mathcal{G}/\mathcal{G}^2$ by

$$x \longmapsto (x-1) + \mathcal{G}^2 .$$

Now θ is a homomorphism, for

$$(xy-1) - (x-1) - (y-1) = (x-1)(y-1) \in \mathcal{G}^2 .$$

Since $\mathcal{G}/\mathcal{G}^2$ is abelian, $G' \subset \ker \theta$, and so θ induces a map $\overline{\theta} : G/G' \to \mathcal{G}/\mathcal{G}^2$.

We construct an inverse to $\overline{\theta}$. Recall (Theorem 5.16) that, as an abelian group, $\mathcal{G}$ is free on all $x-1$, $x \in G$. Define $\phi : \mathcal{G} \to G/G'$ by $\phi(x-1) = xG'$. It suffices to show $\mathcal{G}^2 \subset \ker \phi$, for then ϕ induces a map $\mathcal{G}/\mathcal{G}^2 \to G/G'$ that is visibly the inverse of $\overline{\theta}$.

If $u \in \mathcal{G}^2$, then

$$\begin{aligned} u &= (\Sigma\, m_i(x_i - 1))(\Sigma\, n_j(y_j - 1)) \\ &= \sum_{i,j} m_i n_j (x_i - 1)(y_j - 1) \\ &= \sum_{i,j} m_i n_j [(x_i y_j - 1) - (x_i - 1) - (y_j - 1)] . \end{aligned}$$

Therefore $\phi u = \Pi_{i,j} [x_i y_j x_i^{-1} y_j^{-1}]^{m_i n_j} G'$, and $u \in \ker \phi$, as desired.

COROLLARY 10.3. *For any group* G, $H_1(G, Z) \cong G/G'$.

Exercise 10.1. If p is prime, let Z_p be the cyclic group of order p considered as a trivial G-module. Prove that $H_1(G, Z_p) \cong G/G'G^p$, where G^p is the subgroup of G generated by all p-th powers.

Our next aim is to prove $H^2(G, A) \cong e(G, A)$, where $e(G, A) =$ factor sets/coboundaries. We remind the reader of the formulas.

A factor set is a function $[\ ]: G \times G \to A$ with

(i) $[x, 1] = O = [1, x]$;

(ii) $x[y, z] - [xy, z] + [x, yz] - [x, y] = O$;

a coboundary is a function $[\ ]: G \times G \to A$ with

(iii) $[x, y] = x\langle y\rangle - \langle xy\rangle + \langle x\rangle$,

where $\langle\ \rangle: G \to A$ is a function with

(iv) $\langle 1\rangle = O$.

We had almost established the desired isomorphism in §5. The gap was that the cocycles described there satisfied condition (ii) but not (i); the coboundaries satisfied (iii) but not (iv). There is no choice but to write down some G-projective resolutions of Z. Before taking the plunge, we alert the reader that a topological interpretation of cohomology groups will appear.

DEFINITION. Let $G^{(n)}$ denote the cartesian product of n copies of G; let P_n be the free abelian group on $G^{(n+1)}$ made into a G-module by $x(x_o, x_1, \ldots, x_n) = (xx_o, xx_1, \ldots, xx_n)$. Define maps $\partial_n : P_n \to P_{n-1}$, whenever $n \geq 1$, by

$$(x_o, \ldots, x_n) \longmapsto \sum_{i=o}^{n} (-1)^i (x_o, \ldots, \hat{x}_i, \ldots, x_n) ,$$

where $\hat{}$ indicates deletion.

THEOREM 10.4. *$\mathbf{P} = \cdots \to P_1 \xrightarrow{\partial_1} P_o \xrightarrow{\varepsilon} Z \to 0$ is a G-free resolution of Z.*

Proof: Observe that $P_o = ZG$; ε is, by definition, the augmentation map. We let the reader check that P_n is G-free (a basis consists of all $(n+1)$-tuples of the form $(1, x_1, \ldots, x_n)$). Next, $\mathbf{P}$ is a complex, for the boundary homomorphism is precisely that arising in topology (Theorem 1.3). To prove exactness of $\mathbf{P}$, it suffices to construct a contracting homotopy (exercise 6.8); we want maps

$$\cdots \longleftarrow P_2 \xleftarrow{s_1} P_1 \xleftarrow{s_o} P_o \xleftarrow{s_{-1}} Z$$

with

$$\partial_{n+1} s_n + s_{n-1}\partial_n = 1_{P_n}, \qquad n \geq 0 \quad (\partial_o = \varepsilon)$$

and

$$\varepsilon s_{-1} = 1_Z .$$

Define $s_{-1}: Z \to P_o$ by $1 \mapsto (1)$ (the 1 in parentheses is the identity of G); if $n \geq 0$, define s_n by

$$(x_o, \ldots, x_n) \longmapsto (1, x_o, \ldots, x_n) .$$

Note that the maps s_n are only Z-maps (which is all we need: the conclusion is that $\mathbf{P}$ is exact as a complex of abelian groups; but, by the definition of exactness, $\mathbf{P}$ is then exact as a complex of G-modules). Here are the simple computations.

(i) $\varepsilon s_{-1}(1) = \varepsilon(1) = 1$;

(ii) $(\partial_{n+1}s_n + s_{n-1}\partial_n)(x_o, \ldots, x_n)$

$$= \partial_{n+1}(1, x_o, \ldots, x_n) + s_{n-1} \sum_{i=o}^{n} (-1)^i (x_o, \ldots, \hat{x}_i, \ldots, x_n)$$

$$= (x_o, \ldots, x_n) + \sum_{k=o}^{n} (-1)^{k+1}(1, x_o, \ldots, \hat{x}_k, \ldots, x_n)$$

$$+ \sum_{i=o}^{n} (-1)^{i}(1, x_o, \ldots, \hat{x}_i, \ldots, x_n)$$

$$= (x_o, \ldots, x_n) .$$

Exercise 10.2. Prove that P_n is isomorphic (as a G-module) to the tensor product (over Z) of ZG with itself $n+1$ times.

The resolution **P** of "homogeneous" $(n+1)$-tuples is quite simple, but it does not resemble the formulas that arose in §5. We now construct a resolution of "inhomogeneous" n-tuples.

DEFINITION. For $n > O$, define Q_n as the free G-module with basis all n-tuples $[x_1, \ldots, x_n]$ of elements of G; define Q_o as the free G-module on the single generator [ ].

For each $n \geq O$, $P_n \cong Q_n$, for both are free G-modules on a set in one-to-one correspondence with $G^{(n)}$; we give a specific isomorphism.

Define $\tau: P_n \to Q_n$ by

$$\tau(x_o, \ldots, x_n) = x_o[x_o^{-1}x_1, x_1^{-1}x_2, \ldots, x_{n-1}^{-1}x_n] ;$$

define $\sigma: Q_n \to P_n$ by

$$\sigma[x_1, \ldots, x_n] = (1, x_1, x_1x_2, x_1x_2x_3, \ldots, x_1x_2 \cdots x_n) .$$

It is easy to check that σ and τ are inverse to one another. There are thus unique maps $d_n : Q_n \to Q_{n-1}$, $n \geq 1$, making the following diagrams commute:

$$\begin{array}{ccc} P_n & \xrightarrow{\tau} & Q_n \\ \Big\downarrow \partial_n & & \Big\downarrow d_n \\ P_{n-1} & \xrightarrow{\tau} & Q_{n-1} \end{array} .$$

Since $\tau^{-1} = \sigma$, we have $d_n = \tau \partial_n \sigma$. The formula is

$$d_n[x_1, \ldots, x_n] = x_1[x_2, x_3, \ldots, x_n] + \sum_{i=1}^{n-1} (-1)^i [x_1, \ldots, x_i x_{i+1}, \ldots, x_n]$$
$$+ (-1)^n [x_1, \ldots, x_{n-1}].$$

In particular, we have

$$d_1[x] = x[\ ] - [\ ],$$
$$d_2[x, y] = x[y] - [xy] + [x],$$
$$d_3[x, y, z] = x[y, z] - [xy, z] + [x, yz] - [x, y].$$

Thus $\mathbf{Q}$ is the resolution of Z we began building in §5.

THEOREM 10.5. *$\mathbf{Q} = \cdots \to Q_1 \xrightarrow{d_1} Q_o \xrightarrow{\varepsilon} Z \to O$ is a G-free resolution of Z.*

Proof: First, $\mathbf{Q}$ is a complex, for $d_{n-1} d_n = \tau \partial_{n-1} \partial_n \sigma = O$ and $\varepsilon d_1 = O$. Next, τ: $\mathbf{P} \to \mathbf{Q}$ is an isomorphism of complexes, hence it induces isomorphisms $H_n(\mathbf{P}) \to H_n(\mathbf{Q})$, all n. Since $\mathbf{P}$ is exact, each $H_n(\mathbf{P}) = O$; therefore each $H_n(\mathbf{Q}) = O$ and $\mathbf{Q}$ is exact.

The complex $\mathbf{P}$ suggests a connection between group cohomology and cohomology of topological spaces.

DEFINITION. Let X be a topological space, and let $\mathrm{Aut}(X)$ be the group of all homeomorphisms of X with itself. A group G **operates** on X if there is a homomorphism $G \to \text{Aut}(X)$.

If G operates on X, we may regard each $g \in G$ as a homeomorphism of X; moreover, there are identities

$$g_1(g_2 x) = (g_1 g_2) x \ ,$$
$$1x = x \ ,$$

where $g_1, g_2, 1 \in G$ and $x \in X$.

DEFINITION. A group G operates **without fixed points** on a space X if $gx = x$ for some $x \in X$ implies $g = 1$.

THEOREM 10.6. *If a group G operates on a topological space X, then the singular complex* $\cdots \to S_n(X) \xrightarrow{\partial} S_{n-1}(X) \to \cdots$ *is a complex of G-modules. Moreover, if G operates without fixed points, each* $S_n(X)$ *is a free G-module.*

Proof: Recall that $S_n(X)$ is the free abelian group with basis all continuous $T: \Delta_n \to X$. If $g \in G$, we may regard g as a homeomorphism $X \to X$, so that gT is again an n-simplex. The identities above show that $S_n(X)$ is a G-module. Also (using the notation of §1), if $T^i = Te_i$ is the i-th face of T, then $(gT)^i = gTe_i = g(T^i)$; it follows that ∂ is a G-map.

If $x \in X$, its *orbit* is the set $\{gx: g \in G\}$; clearly the orbits partition X. Let $X_o \subset X$ consist of one element from each orbit. Call a simplex T *basic* if $Tv_o \in X_o$ (where $\Delta_n = [v_o, \ldots, v_n]$); we claim $S_n(X)$ is free on all basic simplexes. They do generate, for if $S: \Delta_n \to X$, then $Sv_o = gx_o$ for some $g \in G$, $x_o \in X_o$; hence $g^{-1}S$ is basic and $S = g(g^{-1}S)$. Finally, suppose $\Sigma\, \gamma_i T_i = O$, where $\gamma_i = \Sigma\, m_{ik} g_k \in ZG$, and the T_i are distinct. Then

$$\sum_{i,k} m_{ik}(g_k T_i) = O .$$

We claim that if $gT_1 = hT_2$, where T_1, T_2 are basic, then $g = h$ and $T_1 = T_2$. Suppose $T_1 v_o \neq T_2 v_o$; then, as these elements lie in distinct orbits, $gT_1 v_o \neq hT_2 v_o$; therefore $T_1 v_o = T_2 v_o = x$. But $gx = hx$ implies $g^{-1}h$ fixes x; since G operates without fixed points, $g = h$. Finally, since g is a homeomorphism, $T_1 = T_2$. It follows that all the simplexes $g_k T_i$ are distinct, so that each $m_{ik} = O$, hence each $\gamma_i = O$.

DEFINITION. A topological space X is **acyclic** if $H_o(X) \cong Z$ and $H_n(X) = O$ for $n > O$.

THEOREM 10.7. *If a group G operates without fixed points on an acyclic space X, then the singular complex of X is a deleted G-free resolution of Z.*

Proof: We already know the singular complex is a complex of G-free modules. The condition that X is acyclic gives an exact sequence $\cdots \to S_1(X) \to S_o(X) \to H_o(X) \to O$, and $H_o(X) \cong Z$.

Let A be a G-module. Now $H^n(X; A)$ is the homology of the complex $\mathrm{Hom}_Z(S(X), A)$. If X is acyclic and G operates on X without fixed points, then $H^n(G, A)$ is the homology of the complex $\mathrm{Hom}_G(S(X), A)$, by Theorem 10.7. To complete this discussion, we assume some knowledge of topology. Suppose G operates *properly* on X, i.e., each $x \in X$ lies in an open set U with $gU \cap U = \phi$, all $g \in G$, $g \neq 1$ (this implies that G operates without fixed points). One can give an isomorphism of complexes (see Mac Lane, p. 135-136)

$$\mathrm{Hom}_Z(S(X/G), A) \cong \mathrm{Hom}_G(S(X), A)$$

where A is G-trivial and X/G is the orbit space of X, and this induces isomorphisms $H^n(X/G; A) \cong H^n(G, A)$.

Suppose Y is an arcwise connected space with fundamental group $\pi_1(Y) = \pi$. If Y has a universal covering space X, then π operates properly on X and X/π is homeomorphic to Y. If, in addition, Y is *aspherical* (i.e., $\pi_n(Y) = 0$, $n > 1$), then X is acyclic. Therefore, if Y is an aspherical space having a universal covering space, then $H^n(Y; A) \cong H^n(\pi_1(Y), A)$ when A is $\pi_1(Y)$-trivial. Such a space Y is called an **Eilenberg-Mac Lane space** $K(\pi, 1)$.

We return to algebra, still seeking to prove $H^2(G, A) \cong e(G, A)$.

DEFINITION. The **bar resolution** (or *standard resolution* or *normalized resolution*) is

$$\mathbf{B} = \cdots \longrightarrow B_1 \xrightarrow{d_1} B_o \xrightarrow{\varepsilon} Z \longrightarrow 0$$

where B_n is the free G-module on all $[x_1, \ldots, x_n]$ with $x_i \in G$ and $x_i \neq 1$, and the formula for $d_n : B_n \to B_{n-1}$ is the same as in $\mathbf{Q}$. (B_o is free on the single generator [ ].)

REMARKS. 1. In order that d_n be defined, we agree that $[x_1, \ldots, x_n] = 0$ whenever some $x_i = 1$.

2. It is not obvious that $\mathbf{B}$ is a complex (this does not follow immediately from $d_{n-1}d_n = 0$ in $\mathbf{Q}$, for here we are making some of the terms in the formula equal to 0).

3. $\mathbf{B}$ is called the bar resolution because the original notation for $[x_1, \ldots, x_n]$ was $[x_1 | x_2 | \cdots | x_n]$.

THEOREM 10.8. *The bar resolution* $\mathbf{B}$ *is a G-free resolution of* Z.

Proof: Again, we do not yet know $\mathbf{B}$ is even a complex (so far, all we know is that each B_n is G-free and the d's are G-maps).

First, we construct a contracting homotopy

$$\cdots \longleftarrow B_1 \xleftarrow{s_o} B_o \xleftarrow{s_{-1}} Z,$$

where each s_n is a Z-map. Define

$$s_{-1}: Z \longrightarrow B_o \text{ by } 1 \longmapsto [\ ]$$

and

$$s_n: B_n \longrightarrow B_{n+1} \text{ by } x[x_1, \ldots, x_n] \longmapsto [x, x_1, \ldots, x_n]$$

(since s_n is only a Z-map, it must be defined on a Z-free set of generators, not merely on the G-free generators $[x_1, \ldots, x_n]$). It is easy to check that this is a contracting homotopy, i.e.,

$$d_{n+1}s_n + s_{n-1}d_n = 1_{B_n} \quad (\text{where } d_o = \varepsilon)$$

and

$$\varepsilon s_{-1} = 1_Z .$$

If we show **B** is a complex, then exercise 6.8 completes the proof. Now B_{n+1} is generated as a G-module by the subgroup im s_n, so that it suffices to show $d_n d_{n+1} = O$ on this subgroup. We do an induction on n, noting that $0 = \varepsilon d_1 = s_{-1}d_1$. For the inductive step,

$$\begin{aligned} d_n d_{n+1} s_n &= d_n(1 - s_{n-1}d_n) && \text{(contracting homotopy)} \\ &= d_n - (1 - s_{n-2}d_{n-1})d_n && \text{(ditto)} \\ &= d_n - d_n - s_{n-2}d_{n-1}d_n \\ &= O . \end{aligned}$$

THEOREM 10.9. $H^2(G, A) \cong e(G, A)$.

Proof: Applying $\mathrm{Hom}_G(\ \ , A)$ to **B** gives the complex

$$\cdots \longrightarrow \mathrm{Hom}_G(B_1, A) \xrightarrow{d_2^*} \mathrm{Hom}_G(B_2, A) \xrightarrow{d_3^*} \mathrm{Hom}_G(B_3, A) \longrightarrow \cdots$$

and

$$H^2(G, A) = \ker d_3^*/\mathrm{im}\ d_2^* \ .$$

Suppose $f\colon B_2 \to A$ is a cocycle. Then $O = d_3^* f = f d_3$, so that

$$O = fd_3[x, y, z] = xf(y, z) - f(xy, z) + f(x, yz) - f(x, y)$$

(that f is a G-map was used to get $fx[y, z] = xf(y, z)$). Moreover, $f(x, 1) = O = f(1, x)$ since we have normalized, so that f is a factor set.

Suppose $g\colon B_2 \to A$ lies in im d_2^*, i.e., there is a map $h\colon B_1 \to A$ with $g = d_2^* h = hd_2$. Since $h\colon B_1 \to A$, we have $h1 = O$. Also, $g(x, y) = hd_2[x, y] = xh(y) - h(xy) + h(x)$ (we used h is a G-map to get $hx[y] = xh(y)$).

THEOREM 10.10. *If G is a finite group of order m, then*

$$mH^n(G, A) = O$$

for all G-modules A and all $n > O$.

Proof: We use the (unnormalized) resolution **Q**. If $f\colon Q_n \to A$, define $g\colon Q_{n-1} \to A$ by

$$g(x_1, \ldots, x_{n-1}) = \sum_{x \in G} f(x_1, \ldots, x_{n-1}, x) \ .$$

Now sum the coboundary formula

$$(df)(x_1, \ldots, x_{n+1}) = x_1 f(x_2, \ldots, x) + \sum_{i=1}^{n-2} (-1)^i f(x_1, \ldots, x_i x_{i+1}, \ldots, x)$$
$$+ (-1)^{n-1} f(x_1, \ldots, x_n x) + (-1)^n f(x_1, \ldots, x_n)$$

over all $x_{n+1} = x \in G$. In the next to last term, as x varies over G, so does $x_n x$. Therefore, if f is a cocycle, $df = O$ and

$$O = x_1 g(x_2, \ldots, x_n) + \sum_{i=1}^{n-2} (-1)^i g(x_1, \ldots, x_i x_{i+1}, \ldots, x_n)$$

$$+ (-1)^{n-1} g(x_1, \ldots, x_{n-1}) + m(-1)^n f(x_1, \ldots, x_n)$$

(the last term is independent of x). Hence

$$O = dg + m(-1)^n f,$$

i.e., mf is a coboundary.

THEOREM 10.11. (Schur-Zassenhaus) *Let E be a finite group having a normal abelian subgroup A with $(|A|, |E/A|) = 1$. Then E is a semidirect product of A by E/A and any two complements of A in E are conjugate.*

Proof: Let $E/A = G$ have order m, and let $|A| = k$. Consider A as a G-module via conjugation (Theorem 5.2). If $n > O$, then $mH^n(G, A) = O$, by Theorem 10.10; also, $kH^n(G, A) = O$, by exercise 7.10, for $kA = O$. Since $(m, k) = 1$, $H^n(G, A) = O$. When $n = 2$, this says that E is a semidirect product of A by G; when $n = 1$, this says that any two complements of A are conjugate (Theorem 5.12).

REMARK. The assumption that A is abelian can be removed.

THEOREM 10.12. *Let G be a finite cyclic group of order k with generator x; let $D = x - 1$ and $N = 1 + x + x^2 + \cdots + x^{k-1} \in ZG$. Then the following is a G-free resolution of Z:*

$$\cdots \to ZG \xrightarrow{D} ZG \xrightarrow{N} ZG \xrightarrow{D} ZG \xrightarrow{\varepsilon} Z \to O,$$

where D, N denotes multiplication by D, N respectively.

Proof: First, ZG is commutative, so that multiplications are G-maps. Clearly $ND = DN = x^k - 1 = O$, so we do have a complex.

Now $\ker \varepsilon = \mathcal{G} = \operatorname{im} D$, by Theorem 5.16, so we have exactness at the first step.

Suppose $u = \Sigma_{i=o}^{k-1} m_i x^i \in \ker D$. Computing gives $m_o = m_1 = \cdots = m_{k-1}$, so that $u = m_o N \in \operatorname{im} N$.

Finally, if $u \in \ker N$, then $O = uN = \Sigma_{j=o}^{k-1} (\Sigma_{i=o}^{k-1} m_i) x^j$, so that $\Sigma\, m_i = O$ and

$$u = -D[m_o 1 + (m_o + m_1)x + \cdots + (m_o + \cdots + m_{k-1})x^{k-1}] \in \operatorname{im} D.$$

COROLLARY 10.13. *Let G be a finite cyclic group of order k. If A is a G-module, let* ${}_N A = \{a \in A : Na = O\}$. *Then*

$$H^o(G, A) = A^G$$

$$\left.\begin{aligned} H^{2n-1}(G, A) &= {}_N A/DA \\ H^{2n}(G, A) &= A^G/NA \end{aligned}\right\} \quad n \geq 1$$

Proof: Apply $\operatorname{Hom}_G(\ , A)$ to the resolution of Theorem 10.12.

Exercises 10.3. If $G = \{1\}$, then, for all A,

$$H^n(G, A) = O = H_n(G, A), \quad n > O,$$

and

$$H^o(G, A) = A = H_o(G, A).$$

(Note that here $ZG \cong Z$, so that G-module means abelian group.)

10.4. If G is finite cyclic, prove

$$H_n(G, A) \cong H^{n+1}(G, A), \text{ all } A, \text{ all } n \geq 1.$$

10.5. If G is finite cyclic, then $hd(Z) = \infty$, where, as usual, Z is a trivial G-module.

10.6 If $R = Z[x]/(x^k - 1)$, then $D(R) = \infty$.

10.7 If G is finite and A is a f.g. G-module, then $H^n(G, A)$ and $H_n(G, A)$ are finite, all n.

10.8 Define $\widetilde{H}^n(G, A) = \operatorname{Ext}^n_{ZG}(\mathfrak{G}, A)$. Prove that

$$\widetilde{H}^0(G, A) \cong \operatorname{Der}(G, A)$$

and

$$\widetilde{H}^n(G, A) \cong H^{n+1}(G, A), \quad n \geq 1.$$

(Hint: use Theorem 5.18 and exercise 5.13.)

THEOREM 10.14. *Let* G *be a free group (nonabelian!) with basis* X. *Then* $\mathfrak{G}$ *is a free G-module with basis all* $x - 1$, $x \in X$.

REMARK. We know from Theorem 5.16 that $\mathfrak{G}$ is Z-free on all $g - 1$, $g \in G$.

Proof: First we show $\mathfrak{G}$ is generated by all $x - 1$. Now

$$xy - 1 = (x - 1) + x(y - 1)$$

and

$$x^{-1} - 1 = -x^{-1}(x - 1),$$

so that if w is a word in X, *then* $w - 1$ can be written as a G-linear combination of $x - 1$, $x \in X$.

To show $\mathfrak{G}$ is freely generated by all $x - 1$, we must complete the diagram

$$\begin{array}{ccc} \mathfrak{G} & \searrow & \\ \cup & & \\ \{x - 1\} & \xrightarrow{\phi} & A, \end{array}$$

where A is a G-module and ϕ is any function. We seek something in $\operatorname{Hom}_G(\mathfrak{G}, A) \cong \operatorname{Der}(G, A)$, so we may as well seek a derivation. Consider the (necessarily) split extension $0 \to A \to E \to G \to 1$, so that E consists of all ordered pairs $(a, g) \in A \times G$. The given function ϕ defines a lifting λ of the generators X of G, namely, $\lambda x = (\phi(x-1), x)$. Since G is free on X, the function $\lambda: X \to E$ extends to a homomorphism $L: G \to E$; moreover, if $g \in G$, then $L(g) = (\langle g \rangle, g)$ for some $\langle g \rangle \in A$. By exercise 5.8, $\langle\,\rangle: G \to A$ is a derivation, so that, by Theorem 5.18, $h: \mathcal{G} \to A$ defined by $h(g-1) = \langle g \rangle$ is a G-map. Since $\lambda(x) = L(x) = (\langle x \rangle, x)$, we have $h(x-1) = \phi(x-1)$ and h extends ϕ.

COROLLARY 10.15. *If* G *is a free group, then*

$$H^n(G, A) = O$$

for all A, *all* $n > 1$.

Proof: By Theorem 10.14, the augmentation sequence

$$O \longrightarrow \mathcal{G} \longrightarrow ZG \longrightarrow Z \longrightarrow O$$

is a free G-resolution of Z.

If S is a subgroup of a group G, then every G-module A is an S-module (for ZS is a subring of ZG).

DEFINITION. A group G has **cohomological dimension** $\leq n$, $cd(G) \leq n$, if

$$H^{n+1}(S, A) = O$$

for all G-modules A, all subgroups $S \subseteq G$. If there is no such integer n, then $cd(G) = \infty$.

EXAMPLES. 1. If $G = \{1\}$, then $cd(G) = O$. This is exercise 10.3.

2. If G is finite cyclic of order $k > 1$, then $cd(G) = \infty$. This follows from Corollary 10.13, taking $A = Z$.

3. If G is free, (and $\neq \{1\}$), then $cd(G) = 1$. By Corollary 10.15, $cd(G) \leq 1$ (for every subgroup of a free group is free). Furthermore, $cd(G) \neq O$, for $H^1(G, Z) \cong \mathrm{Hom}(G, Z)$ since Z is G-trivial.

4. If $cd(G) < \infty$, then G is torsion-free. If G is not torsion-free, it has a finite cyclic subgroup $S \neq \{1\}$, and $H^n(S, Z) \neq O$, all even n.

5. One may show that if G is a free abelian group of finite rank n, then $cd(G) = n$.

What are the groups with $cd(G) = 1$? Let Z_2G denote the group algebra of G with coefficients in $Z/2Z$. Stallings (*Annals of Math.*, 1968) proves the following

THEOREM. *Let G be a finitely presented torsion-free group for which $H^1(G, Z_2G)$ has more than two elements; then G is a non-trivial free product.*

As consequences, he obtains the following results.

COROLLARY. *If G is a finitely generated group with $cd(G) = 1$, then G is free.*

COROLLARY. *If G is a torsion-free, finitely generated group having a free subgroup of finite index, then G is free.*

Swan (1969) has shown that both corollaries are true if one removes the hypothesis that G is finitely generated.

REFERENCES

[1] Cartan, H. and Eilenberg, S., *Homological algebra*, Princeton, 1956.

[2] Curtis, C. and Reiner, I., *Representation theory of groups and associative algebras*, Interscience, 1962.

[3] Eilenberg, S., and Steenrod, N., *Foundations of algebraic topology*, Princeton, 1952.

[4] Fuchs, L., *Abelian groups*, Budapest, 1958.

[5] Godement, R., *Théorie des faisceaux*, Hermann, 1958.

[6] Gruenberg, K., *Some cohomological topics in group theory*, Queen Mary College Math. Notes, London, 1967.

[7] Jans, J., *Rings and homology*, Holt, 1964.

[8] Kaplansky, I., *Commutative rings*, Queen Mary College Math. Notes, London, 1966.

[9] Lambek, J., *Lectures on rings and modules*, Blaisdell, 1966.

[10] MacLane, S., *Homology*, Academic Press, 1963.

[11] Northcott, D., *Introduction to homological algebra*, Cambridge, 1960.

[12] Serre, J.-P., *Corps locaux*, Hermann, 1962.

[13] Zariski, O., and Samuel, P., *Commutative algebra, I.*, van Nostrand, 1958.

INDEX